W9-BQL-129
LENOIR RHYNE COLLEGE

BOOK 5

Cell Membranes

BOOK 5

Cell Membranes

Basic Biology Course

Unit 1   Microscopy and its Application to Biology
Book 1 *Light Microscopy*
Book 2 *Electron Microscopy and Cell Structure*
Book 3 *Dynamic Aspects of Cells*
    Film strip for Book 2
    Tape commentary (cassette) for Book 2
    4 super-8 film loops for Book 3

Unit 2   Organisms and their Environment
Book 4 *Ecological Principles*
The Ecology Game pack
    Film strip for Book 4
    Film strip for Ecology Game

Unit 3   Regulation within Cells
Book 5 *Cell Membranes*
Book 6 *Photosynthesis*
Book 7 *Enzymes*
Book 8 *Metabolism and Mitochondria*
Book 9 *Protein Synthesis*
The Enzyme Game pack
    Super-8 film loop for Book 5
    Film strip for Book 6 and Book 8
    Tape commentary (cassette) for Book 6 and Book 8

Unit 4   Communication between Cells
Book 10 *Nerves*
Book 11 *Hormones*
    Film strip for Book 10
    Film strip for Book 11

Tutors' Guide

BASIC BIOLOGY COURSE
UNIT 3
REGULATION WITHIN CELLS

BOOK 5

Cell Membranes

MICHAEL A. TRIBE, MICHAEL R. ERAUT &
ROGER K. SNOOK
University of Sussex

CAMBRIDGE UNIVERSITY PRESS

CAMBRIDGE
LONDON · NEW YORK · MELBOURNE

Published by the Syndics of the Cambridge University Press
The Pitt Building, Trumpington Street, Cambridge CB2 1RP
Bentley House, 200 Euston Road, London NW1 2DB
32 East 57th Street, New York, NY 10022, USA
296 Beaconsfield Parade, Middle Park, Melbourne 3206, Australia

© Cambridge University Press 1976

Library of Congress catalogue card number: 75–7217

ISBNs:
    0 521 20737 1 hard covers
    0 521 20738 X limp covers

First published 1976

Printed in Great Britain
at the University Printing House, Cambridge
(Euan Phillips, University Printer)

Contents

Foreword   *page* vii
Acknowledgements   viii

5.0 Introduction   1

  5.0.1. Discussion   1
  5.0.2. Overview   1
  5.0.3. Preknowledge requirements   2
  5.0.4. Objectives   2
  5.0.5. Instructions on working through programmed sections   3

5.1. The structure and function of cell membranes   4

  5.1.1. Structure   4
  5.1.2. Function   24

5.2. Phagocytosis and pinocytosis   49

  5.2.1. Instructions for using film loop projector   50
  5.2.2. Phagocytosis   51
  5.2.3. Pinocytosis   61

5.3. Concluding remarks   65

5.4. Glossary of terms used   67

5.5. Appendix. Practical experiments   69

  Expt 1.   Observations on the fate of food vacuoles
            in *Paramecium*   69
  Expt 2.   Observations on contractile vacuole activity
            in relation to the concentration of the external
            medium in *Podophrya* or *Paramecium*   70
  Expt 3.   Observations on the chemotactic behaviour
            of amoebae   72

5.6. Questions relating to the objectives of the book   73

5.7 Recommended reading   78

Index   79

Foreword

This book is part of a Basic Biology Course for undergraduates written by the Nuffield Inter University Biology Teaching Project team at Sussex.

The main aim of the book is to provide you with an insight into the structure and function of cell membranes, sufficient for you to examine critically some of the models proposed for their structure and to enable you to appreciate the vitally important, yet complex role which they play in regulating life processes.

Book 5 is in fact one of five books (Books 5 to 9 inclusive) comprising a unit called 'Regulation within cells' (See the outline of course structure at front of book.) It also provides a necessary background for Books 10 and 11 dealing with aspects of 'Communication between cells', such as conduction of impulses by neurons and hormonal action.

Sussex, 1974

Michael A. Tribe
Michael R. Eraut
Roger K. Snook

Acknowledgements

This book was developed under the auspices of the Inter University Biology Teaching Project and is the responsibility of the Sussex University project team. However, it owes a great deal to the students who studied and criticized our earlier versions and to many colleagues both at Sussex and elsewhere who made constructive suggestions for its improvement.

In particular we would like to thank the following:

Dr K.P. Wheeler, University of Sussex;

Dr I. Tallan, on leave from the University of Toronto (1974–5);

the Nuffield Foundation for financially supporting the project from 1969 to 1972;

Cambridge University Press for the continued interest and support in publishing the materials;

Mrs P. Smith and Mrs S. Collier project secretaries;

Mr C. Atherton for photographic assistance.

We are extremely grateful to the following for allowing us to use their electron micrographs:

Dr J. Beggs, Barrow Neurological Institute (page 56)

Dr D. Branton, Dept. of Botany, University of California, Berkeley (pages 20, 23)

Drs W.H. Butler & J.H. Judah, MRC Toxicology Laboratory, Carshalton (inset, page 14)

Dr D.W. Fawcett, Dept. of Anatomy, Harvard Medical School (pages 62 B & C; 63)

Dr A. Gropp, Pathologisches Institute, University of Bonn, W. Germany (page 60)

Drs Henn, Decker, Greenawalt & Thompson, Johns Hopkins University (page 13)

Dr D. Prescott, Oak Ridge National Laboratory (page 62 A)

Dr J.D. Robertson, Harvard Medical School (page 12)

Dr F. Sjöstrand, University of California, Los Angeles (pages 10, 14)

5.0. Introduction

5.0.1. Discussion

The best starting point for the investigation of any system is an analysis of its input and output. So it is logical to begin a study of the cell with an account of cell membranes. The plasma membrane determines what goes into the cell and what comes out of it; and the cytomembranes determine the input and output of each of the cell's subsystems, such as the nucleus and the mitochondrion. Taken together these cell membranes normally constitute 50 per cent of the dry mass, which remains when a cell is broken up and centrifuged and its 'soluble' cytoplasm poured off; in some instances the membranous fraction may account for 70–80 per cent of the dry mass.

Present-day research into the structure and function of biological membranes is extremely active and uses sophisticated biophysical and biochemical techniques.

The importance of cell membranes, particularly the plasma membrane, in determining selective permeability and controlling the transport of substances into and out of cells is now well established. Recent research, however, is revealing other fascinating roles for the plasma membrane. For example, it is concerned in cell recognition, now an important aspect of cancer research. Whereas normal cells exhibit auto-immune responses and can distinguish 'self' from 'non-self', it appears that cancerous cells do not recognize 'self', and it is possible that differences in the plasma membrane may be responsible. Another role is the provision of receptor sites for hormones or man-made drugs.

Lastly, since the plasma membrane is that part of the cell which comes into contact with other cells or surfaces, it is intimately involved in activities such as cell movement, cell adhesion, cell fusion, phagocytosis, pinocytosis, and secretory processes of various kinds.

5.0.2. Overview

Preceding books in this course were concerned with:
  (1) The variety and fine structure of different cells in living organisms as revealed by light and electron microscopy (Books 1, 2 and 3).
  (2) The dependence of organisms on their environment; the degree of tolerance shown by organisms with respect to changeable environmental conditions; and the pattern of inter-relationships between organisms, culminating in the concept of ecosystems (Book 4).

In this unit (Unit 3), consisting of Books 5 to 9 inclusive, we aim to:

  (1) examine evidence for the function of organelles;
  (2) examine the inter-relationships which exist between the various inclusions of the cell and point out the advantages of cellular organization;
  (3) indicate those factors which tend to limit the size of cells; and
  (4) examine some of the mechanisms by which a cell is able to control its growth and metabolism.

CELL MEMBRANES

5.0.3. Preknowledge requirements

Basic knowledge of electron microscopy and cell structure as presented in
Books 1 and 2 (Unit 1) in this series.

A basic knowledge of the following chemical substances and terms:
gases, liquids, solids; solute, solvent; fat solvents (e.g. chloroform,
benzene, etc); diffusion, osmosis; acids, bases and electrolytes; polar and
non-polar groups. An idea of the approximate molecular size and
chemical composition of sugars, starches, lipids and proteins (stereo-
chemical details or knowledge of secondary and tertiary structures of
proteins is not assumed).

5.0.4. Objectives

At the end of this book students should be able to:
   (1)   Define and give examples of the following terms:
            diffusion, facilitated diffusion, osmosis, active transport.
   (2)   Present models explaining the structure and some of the functions of
         cell membranes as deduced from experimental evidence.
   (3)   Given experimental data, deduce whether transport is active or
         passive.
   (4)   Indicate the effect of lipid solubility and molecular size of non-
         electrolytes on their rate of penetration into cells.
   (5)   Recognize the specific problems of transporting strong electrolytes
         (particularly Na^+, K^+ and Cl^- ions) across cell membranes and
         briefly describe the involvement of Na^+ ions in glucose and amino
         acid transport in the mammalian intestine.
   (6)   Define and explain the terms phagocytosis, pinocytosis, antigen,
         antibody, opsonin, with special reference to blood cells, capillary
         blood vessel cells, fibroblasts and amoebae.

INTRODUCTION

5.0.5. Instructions on working through programmed sections

In the programmed sections, questions and answers are arranged sequentially down the page. You are provided with a masking card and probably a student response booklet. Sections 5.1 and 5.2 of this book are programmed. Cover each page in turn, and move the masking card down to reveal two thin lines.

This marks the end of the first question on that page. Record your answer to the question under the appropriate section heading in the response booklet provided. Then *check* your answer with the answer given. If your answer is correct, move the masking card down the page to the next double line and so on. If any of your answers are incorrect retrace your steps and try to find out why you answered incorrectly. If you are still unable to understand the point of a given question, make a note of it and consult your tutor.
The single thick line

---

is a demarcation between one frame and the next.
A double bold line indicates a convenient stopping point in the programme, since it is unlikely that you will have time to read through the whole book in one session.

5.1. The structure and function of cell membranes

5.1.1. Structure

1   Those of you who read Book 2 (*Electron microscopy and cell structure*) in this series, will recall that fixatives such as osmium tetroxide and potassium permanganate clearly reveal the position of cell membranes by reacting with *the lipids and proteins that form the basis of membrane structure.*

    Another line of enquiry which was begun by Overton at the beginning of the century, looked at the capacity of various substances to penetrate the plasma membrane.

    The following table shows the type of results obtained.

Table 1

		Penetration		
Very rapid	Rapid	Slow	Very slow	Virtually none
Gases: CO_2	Water	Glucose	Strong electrolytes	Proteins
O_2		Amino acids	(e.g. inorganic salts)	
N_2		Glycerol	Acids	Polysaccharides
Alcohol			Bases	Phospholipids
Ether		Fatty acids		
Chloroform			Sucrose	
Benzene			Maltose	
Carbon tetrachloride			Lactose	

You will notice from the table that organic solvents penetrate the membrane at a faster rate than water, and that strong electrolytes penetrate very slowly. What does this suggest about the nature of the plasma membrane?

That it is predominantly non-polar, i.e. lipid, in character

2   What reasons can you give to account for the finding that water, which is more polar than glycerol, penetrates the membrane more rapidly than glycerol?

    (i)   The small size of the water molecule might account for its relatively rapid penetration because small molecules diffuse faster than large ones. This would also explain why monosaccharides like glucose penetrate faster than disaccharides such as sucrose.

4

(ii) The presence of small polar 'pores' in the membrane could account for the rapid penetration of water and other small polar molecules.

---

3 Proteins, polysaccharides and phospholipids are known to exist inside cells but cannot penetrate the cell membrane. So how do they get inside?

They have to be made inside the cell itself from smaller components which can be imported. (Though this is generally true there are a few exceptional mechanisms for importing these compounds, which we will discuss later.)

---

4 This characteristic of the plasma membrane is called selective permeability (see Glossary) and gives the cell some considerable control over its input and output. In what ways might such control be beneficial to the cell?

(i) It prevents sudden changes in the concentration of all but the most rapidly penetrating molecules.

(ii) It keeps many chemicals which are essential to the life of the cell within the boundaries of the cell. (*Note.* If the plasma membrane is torn with a fine needle the contents ooze out.)

(iii) It allows the rapid penetration of gases such as CO_2, O_2, and N_2 which, as we noted in Book 4 (*Ecological principles*), are important factors in the interaction between many organisms and their environment.

---

5 The predominantly non-polar character of membranes separates them quite clearly from their environment, which is aquatic on both sides, and therefore polar. What advantages can you suggest for having internal cytomembranes as well as a plasma membrane?

By creating internal compartmentation they permit a greater degree of organization. This is one of the factors which enables the cell to perform its functions in a manner more similar to that of a production line than that of a cottage industry.

---

6 Let us now examine membranes more closely, beginning with the lipid component.
Overpage you will see two different phospholipid molecules commonly

5

found in biological membranes; namely, a phosphatidyl choline (or lecithin) and a phosphatidyl ethanolamine. Each molecule resembles a two-pronged peg, in which the choline and ethanolamine residues respectively are soluble in water and are called *polar* groups because they carry a positive electric charge. The long chain fatty acid residues (or side chains) are insoluble in water and are often referred to as the *apolar* region of the molecule. Notice too, that the two side chains are often derived from different fatty acid origins. The phosphatidyl choline illustrated, for example, has one side chain derived from palmitic acid (a saturated fatty acid because it contains no double bonds between carbon atoms) and the other side from oleic acid (an unsaturated acid since —C=C— is present).

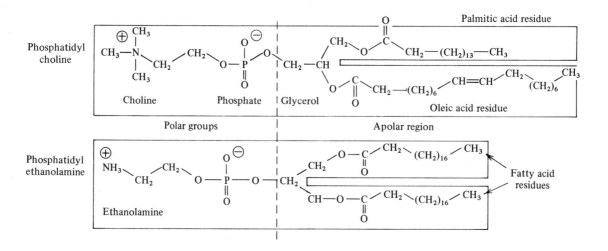

If a thin layer of these molecules is spread over a water surface, how would the molecules orientate themselves with respect to this surface?

With their polar groups (residues) in the water and the insoluble fatty acid residues uppermost.

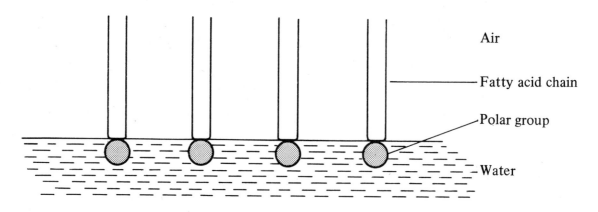

Orientated molecules at an air/water interface form a monolayer or unimolecular layer. When the phospholipid concentration is sufficiently high to more than saturate the interface, micelles are formed, e.g.

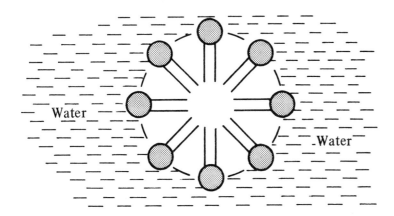

7   What would happen (a) if a little oil is shaken in an excess of water?
                        (b) if a little water is shaken in an excess of oil?

(a)   Little drops of oil would be scattered throughout the water.
(b)   Little drops of water would be scattered throughout the oil.

8   Draw diagrams to illustrate the orientation of phospholipid micelles when
        (a)   a little phospholipid is present in an excess of water
        (b)   a little water is present in an excess of phospholipid.
     (*Hint.* Use the answers to frames 6 and 7)

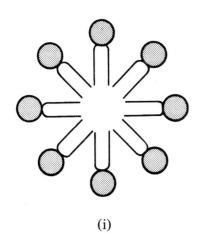

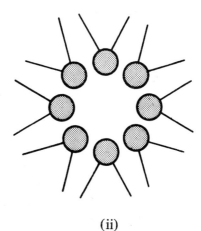

(i)   (ii)

9   In 1925, Gorter & Grendel tried to estimate the thickness of the red blood cell membrane. They took a known number of cells (with defined surface area) and extracted the lipids from the red cell plasma membrane. The extracted lipid was then spread over a water surface. The area of the coherent unimolecular film (which was approx. 2 nm thick) proved to be twice as large as the total surface area of the red blood cells from which it was derived.

What explanation can you give for this result?

_____

_____

The result suggests that the lipid component of the plasma membrane consisted essentially of a lipid bilayer (bimolecular film) about 4 nm thick.

Note. Although Gorter & Grendel made the correct deductions from their experimental data, they actually made an error in their calculations of cell surface area, but this error was cancelled out because (unknown to them) they also failed to extract all the lipids in the red blood cell membranes because the extraction techniques at that time were inadequate. A remarkable coincidence in Science!

10   The other important components of the plasma membrane are proteins. Evidence for their presence is inferred from electron microscope studies together with biochemical analysis.

Our evidence so far indicates, therefore, that a lipid bilayer is present together with protein. The plasma membrane is in other words a lipoprotein complex. The question we need to ask is how the protein is arranged in relation to the lipid.

One such model we could propose is as follows:

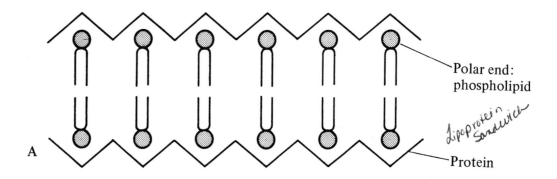

A

Polar end: phospholipid

lipoprotein sandwich

Protein

Here the polar lipid heads are linked to the outer protein coat by ionic and/or hydrogen bonds. What other arrangements of the protein in relation to the phospholipid are possible?

_____

_____

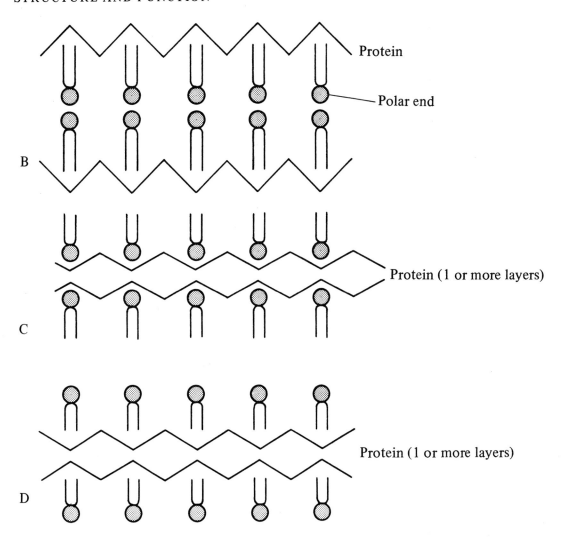

11 We mentioned earlier that the environment both within and outside the cell is essentially aqueous. Which of the above models does this piece of evidence rule out?

C. It is unlikely that the non-polar groups of the phospholipids would project into an aqueous medium.

12 Below is an electron micrograph of two adjacent cell membranes
(marked PM and arrowed) ✕ 300 000 magnification.

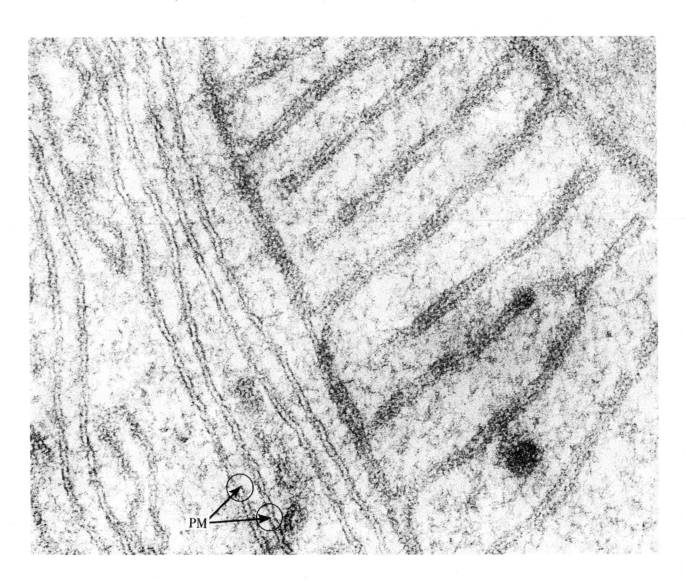

Each membrane (ringed PM) is approximately 7.5 nm thick and would
not, therefore, be visible under the light microscope.

The fixative and staining used here reveals the protein as dark bands.
Most electron microscopic evidence from work with isolated lipid
bilayers suggests that the polar ends of the phospholipid molecules
become darkly stained in preference to the non-polar fatty acid
residues. On the basis of this evidence refer to the models A, B, C, D
proposed earlier (frame 10).

Which model best interprets the electron microscopic evidence?

_____

_____

Model A, in which the non-polar portions of the phospholipid molecules
are directed inwards and the lipid bilayer is stabilized by the adsorption
of monolayers of hydrophilic protein on the outside. If model B or D
were correct we should expect to see three dark bands:

=== Protein

——————————————————————————— Polar ends

=== Protein

If model C were correct (see previous objection) then only one dark band would be seen.

---

13   There is one other feature of importance in animal cells.
Look at the area between the two adjacent cell membranes, it is approximately 15 nm wide. How would you interpret this clearer area? (Seen better on page 12, B.)

(i)   It may represent a fluid-, solid- or gel-filled space.
(ii)   The plasma membranes may be covered with a thick outer surface coat, which does not react with OsO_4 or $KMnO_4$.
(iii) It may be an artefact produced by fixation of the specimen.

---

14   What microscopic evidence could distinguish (i) from (ii)?

The amount of variation in the observed intercellular distance. In fact this is fairly constant at 15–20 nm; and it is now known that the outer layer of protein is coated with a mucopolysaccharide (large complex sugar) forming a mucoprotein or glycoprotein.

---

15   In 1959, Robertson proposed a 'unit membrane' hypothesis to account for the basic structure of all biological membranes,* and the model that he proposed resembled A in frame 10.
     Much of the physico-chemical evidence for the hypothesis was based on data originally collected by Danielli & Davson in 1935, whilst Robertson's own contribution to the theory was based on electron microscopic evidence as shown in A and B below. In obtaining these excellent micrographs Robertson used potassium permanganate as fixative.

*Although Robertson proposed that all biological membranes have a uniform type of structure, with a characteristic trilaminar appearance under the EM, he was *not* saying that all biological membranes are the same.

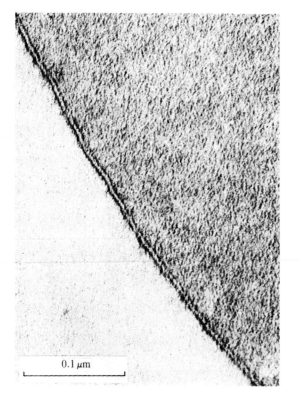

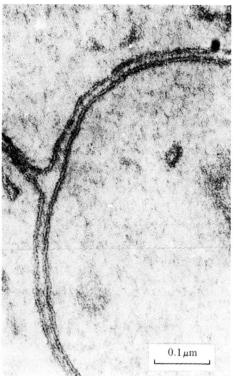

0.1 μm

0.1 μm

A. Part of a human red blood cell.

B. Mature unmyelinated mouse nerve cell.

However, when presented with electron microscopic evidence alone, should you:
(i)   reject it as unsubstantial without other lines of evidence;
(ii) attach considerable importance to it;
(iii) accept it cautiously whilst realizing its limitations?

(iii) is the best answer.
Rejecting any line of evidence, however tenuous and unsupported is dangerous — many of the important discoveries in science have arisen from chance observations.
   In contrast, attaching considerable importance to only one line of evidence, without corroborative evidence from other lines of enquiry is poor scientific practice.

16   With present-day techniques it is possible to make 'artificial' model membranes in water which are known to consist either of a specific lipid bilayer or a lipid bilayer with adsorbed protein.
     How might this model-building technique help you to corroborate the electron microscopic evidence obtained from intact cells?

After fixation and contrast-staining the model membrane should have the same or similar appearance to the natural membranes of cells (see below).

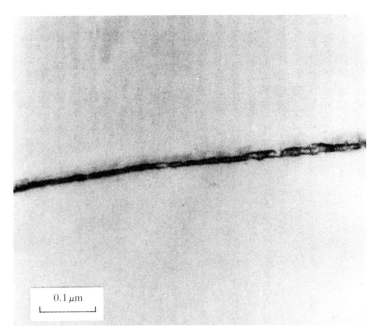

0.1 μm

Electron micrograph of a bilayer of phosphatidyl ethanolamine in *n*-decane fixed with lanthanum nitrate and $KMnO_4$. Reproduced with permission from *J. Mol. Biol* **24**, 51 (1967).

---

17   From your knowledge of fixation techniques (Book 2), however, what objection might there be to the 'unit membrane' theory based on electron micrographs?

Preparation for the electron microscope involves physical and chemical transformation of the membrane; the bilayer structure finally revealed may simply represent a stable configuration of the constituents of the membrane brought about by fixation.

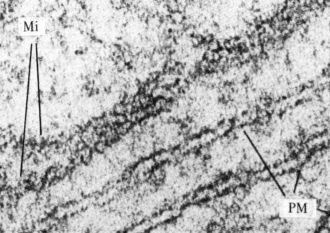

50 nm

Mi

PM

PM

CM

PM

CM

Reproduced with permission from *Int. Rev. Cytol.* **5**, 455 (1956). Inset from *J. Cell Biol.* **44**, 278-89 (1970).

0.1 μm

18   Below you see two diagrams illustrating the appearance of mitochondrial membranes with (a) permanganate fixation and (b) osmium fixation.
What structural differences do you observe?

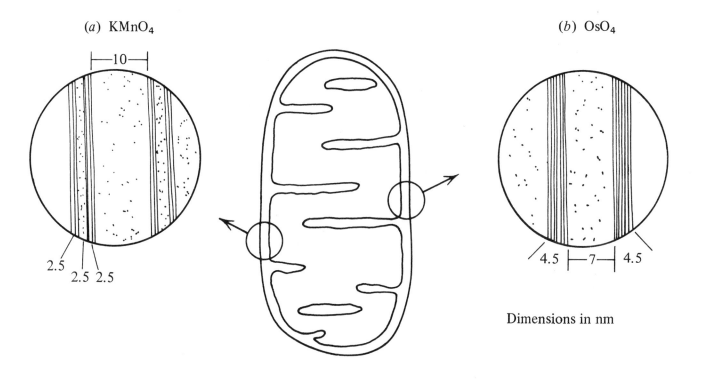

(a) KMnO$_4$

(b) OsO$_4$

Dimensions in nm

There is a difference in appearance of both the inner and the outer membranes.
In (a) each membrane reveals two dense lines separated by an electron-lucid line.
In (b) only two dense lines are seen, each dense line representing a membrane.
It is possible that the two membrane elements are normally fused, but with permanganate fixation split as the mitochondrion swells.
(*Note.* This only happens with mitochondrial membranes, not the plasma membrane, but it emphasizes the effect which can be obtained by different fixation procedures. You might also be interested to know that the three-layered image of the inner mitochondrial membrane remains essentially unchanged even after removal of nearly all lipids!)

19   Now refer to the micrograph opposite. It shows a highly magnified view of the plasma membrane (PM) and the two mitochondrial membranes (Mi). (CM) refers to other cytoplasmic membranes. What differences between the plasma membrane and the mitochondrial membrane do you detect?

15

(i) Each mitochondrial membrane appears slightly thinner (i.e. in total cross-section) than the plasma membrane. In fact when Sjöstrand in 1963 made a number of critical measurements from micrographs of this kind he found a 50% difference between the two types of membrane.

(ii) There is evidence (see inset particularly) that the mitochondrial membrane is 'globular' in structure rather than in a continuous sheet.

One model which has been proposed is as follows:

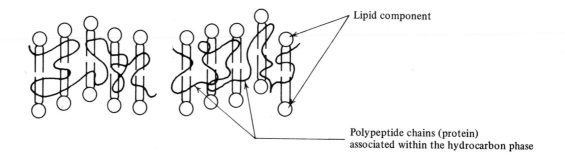

Lipid component

Polypeptide chains (protein) associated within the hydrocarbon phase

Note. The 'unit membrane' model specifies that protein-lipid interactions are polar. However, although there are some polar interactions, much of the physico-chemical evidence indicates that these mechanisms do not dominate the interactions between lipids and proteins (i.e. some of the polypeptide chains are in the hydrocarbon phase).

---

20 Because of this and other evidence which is available, Lucy in 1964 suggested that certain membranes might possess globular micelles of lipid as well as lipid bilayers. These globules may be a prominent feature of some membranes, but might exist in dynamic equilibrium with a bimolecular layer of lipid in other types of membrane.

Lucy's model

Structural protein or glycoprotein

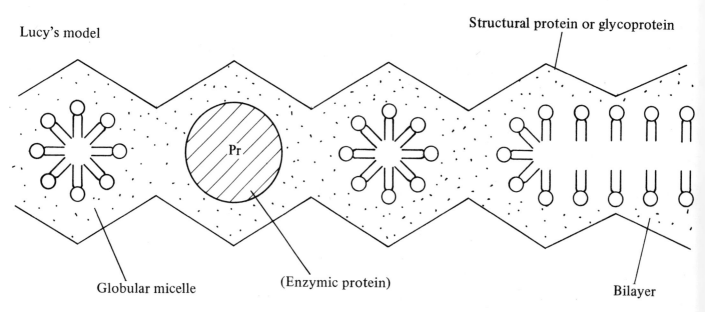

Globular micelle

(Enzymic protein)

Bilayer

16

Such a model would satisfy much of the evidence for a unit membrane and also take account of Sjöstrand's evidence shown on page 14. Also, with this model it is possible to envisage certain positions being occupied by functional (enzymic) rather than structural proteins in more apolar regions of the membrane. Although the function of (enzymic) proteins will become clearer later in this section, can you suggest how functional proteins might increase the membrane's control of molecules entering or leaving the cell?

There are at least two possibilities:
(i)   by inducing physical changes in the structure of the membrane and hence changing its permeability properties;
(ii) by directly controlling the fluxes of molecules chemically by interaction with specific kinds of molecules, either within the membrane or at membrane surfaces.

21   'Flexibility', in both the physical and chemical senses, which will allow the membrane to expand and contract or to add new components, is an essential feature of living cells.
Why?

(i)   It permits the cells to grow.
(ii)   It enables the cell to move.
(iii) It allows the cell to adjust its shape and size as one method of adapting to environmental conditions.
Note. Expansion and contraction, however, are still possible with the 'unit membrane' model and even artificial lipid bilayers are known to be physically flexible and fluid.

A review of the situation

22   So far, we have examined some of the pros and cons of three membrane models (i.e. the unit membrane model, and the models of Sjöstrand and Lucy). Several other models are also possible and have been proposed by various authors in accordance with different lines of experimental evidence; yet none of the models are totally compatible with all the evidence which has been obtained from different membranes. Tables 2 and 3 overpage summarize some of the measurements which have been obtained with various cell membranes and with artificial lipid bilayers.

CELL MEMBRANES

When you look at the tables below, you will not be surprised to find that no single model can encompass all the experimental data from different membranes. The spatial arrangement of the lipids and proteins comprising cell membranes is in many cases complex and probably dynamic (i.e. subject to alterations with conditions and time). Therefore, the argument whether a phospholipid bilayer exists or does not exist; whether the associated protein is in a layer (β–sheet) or globular (α–helical) form; whether the protein is in the apolar or the polar environment of the membrane, almost certain depends on (a) the membrane under consideration, and (b) the physical and chemical conditions under which the experimental evidence was obtained.

Table 2 *Protein and lipid content of membranes* (Adapted after Wallach, 1972)

Membrane type	Protein : lipid content (wt/wt)	Cholesterol : polar lipid content (mol/mol)	Major polar lipids
Myelin sheath around neurons	0.25	0.7 –1.2	Cer, PE, PC
Plasma membranes			
Liver cells	1.0–2.3	0.3 –0.5	PC, PE, PS, Sph
Red blood cells	1.5–4.0	0.8 –1.0	PC, PE, PS, Sph
Endoplasmic reticulum	0.7–1.2	0.03–0.08	PC, PE, Sph, Plas
Mitochondria			
Outer membrane	1.2	0.03–0.09	
Inner membrane	3.6	0.02–0.03	
Retina rod cells of the eye	1.5	0.13	PC, PE, PS
Chloroplast lamellae	0.8	0	Gal DG, SL, PS

PE = Phosphatidyl ethanolamine	Sph	= Sphingomyelin
PC = Phosphatidyl choline	Cer	= Cerebrosides
PS = Phosphatidyl serine	Plas	= Plasmalogen
SL = Sulpholipid	Gal DG	= Galactosyldiglyceride

Do *not* try to remember these names!

Table 3 *Some physical properties of natural and artificial membranes*

Physical property	Natural membranes (range obtained from different membranes)	Artificial membranes (various phospholipids)
Thickness (nm)	5–12	6.8 –7.3
Surface tension (dynes/cm)	0.03–3.0	0.5 –1.0
Permeability to water ($cm/s \times 10^3$)	0.03–3.3	0.5 –1.0
Electrical capacitance ($\mu F/cm^2$)	0.5 –1.3	0.33–1.3
Electrical resistance (ohms/cm^2)	$4 \times 10^2 - 1 \times 10^4$	$10^4 - 10^8$

For example, the myelin sheath ('insulating wrapping') surrounding many nerve cells may conform quite closely to the unit membrane model. However, the mitochondrial inner membrane or the chloroplast lamellae (see Book 2) have to carry out complex multi-enzymic functions associated with oxidative phosphorylation (Book 8) and photophosphorylation (Book 6) respectively. It is hardly surprising to find that these membranes (see Table 2) have a high protein : lipid content (particularly globular, enzymic proteins), or to learn that some of these enzymes (e.g. ATPases) are thought to operate in an apolar (hydrophobic) environment. These functions are certainly consistent with the biophysical evidence obtained from sophisticated optical techniques applied to various cell membranes. It also agrees with the evidence from electron micrographs of freeze-etched or freeze-fractured membranes (Book 2), where the freeze-fracturing process is thought frequently to split membranes at the hydrophobic (apolar) centre:

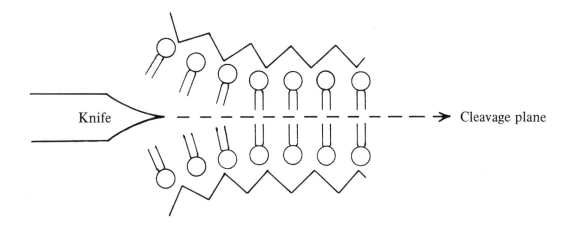

Now look at the electron micrographs on the next page showing freeze-etched preparations obtained from: (A) concentric layers of a myelin sheath, where the surfaces are smooth and devoid of membrane-associated particles; (B) human red blood cell membrane; and (C) part of the granal lamellae from a chloroplast. In (B) and particularly (C) complex membrane-associated particles are present.

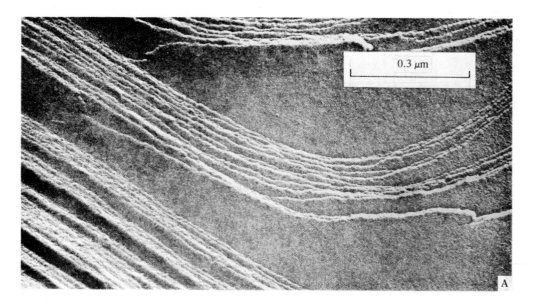

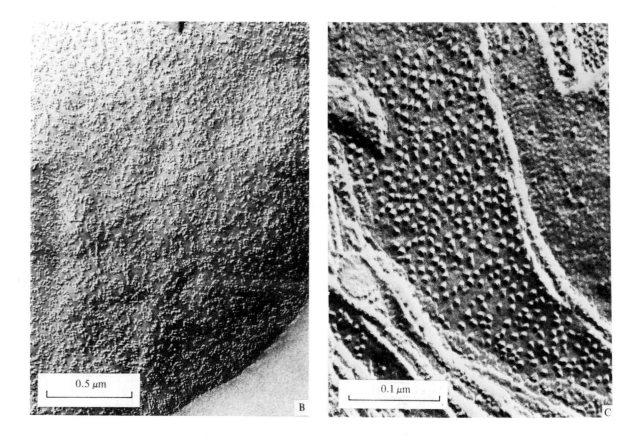

A. Reproduced with permission from *Experimental Cell Res.* **45**, 703–7 (1967).
B. Reproduced with permission of the Royal Society from *Phil. Trans. Roy. Soc.* B, **261**, 133 (1971).

STRUCTURE AND FUNCTION

From estimations of particle density observed in various freeze-etched
membranes under the electron microscope, it would seem that the
most metabolically active membranes have the greatest frequency of
particles, whereas in metabolically inert membranes, membrane-
associated particles are lacking, as shown in Table 4 below.

Table 4

Type of membrane	Percentage of membrane covered by particles
Plasma membranes	
onion root tip	15
human red blood cell	23
yeast cell	63
Myelin sheath	0
Endoplasmic reticulum	
onion root tip	12
Nuclear membrane	
onion root tip	12
Chloroplast lamellae	80
Mitochondrial inner membrane	
mouse intestinal cell	75

Other evidence suggests that:
(1) some areas of the membrane have 'naked' areas of phospholipid
(probably quite extensive in some membranes);
(2) in many membranes the proteins are distributed fairly randomly,
although in others they may be ordered and even show polarity;
(3) the protein component of the membrane is found associated with
both apolar and polar regions; and
(4) freeze-etched membranes under the electron microscope show in
most cases membrane-associated particles on the inside surface,
outside surface and internal hydrophobic surfaces.
Perhaps our membrane model should look something like this:

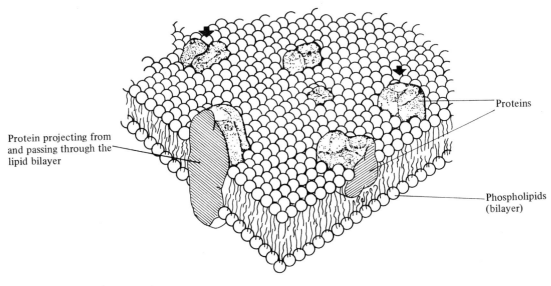

Protein projecting from
and passing through the
lipid bilayer

Proteins

Phospholipids
(bilayer)

Redrawn after Finean (1972)

21

If the particles in the freeze-etched electron micrographs are the proteins postulated in this model, what likely effect would treating a membrane with a proteolytic enzyme (e.g. pronase) have?

It would be expected to remove many, if not all the particles from freeze-etched membranes depending on the time of exposure to the enzyme. (This has been confirmed by Branton with the plasma membrane of red blood cells, as shown in the sequence of pictures on the opposite page.)

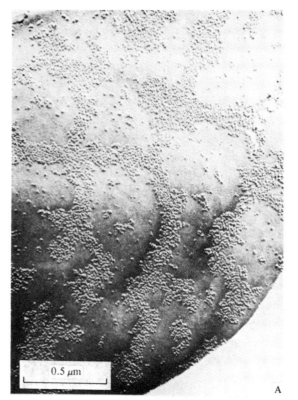

0.5 μm

A

0.5 μm

B

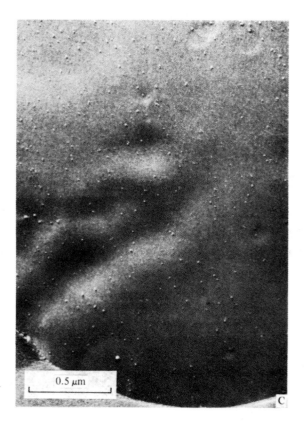

0.5 μm

C

The effects of pronase on red blood cell plasma membrane. A, B and C are the results of incubating the material in buffer + pronase for varying periods of time. A — 30% particles removed; B — 45%; C — 70%. Reproduced with permission of the Royal Society from *Phil. Trans. Roy. Soc.* B, **261**, 133 (1971).

5.1.2. Function

23   *Do 'pores' exist in the plasma membrane?*
In attempting to correlate membrane structure with function, we come up against a problem. From Overton's evidence (Table 1) we can see that cell membranes exhibit certain filter effects, whereby small molecules (actually substances with molecular weights less than 80) generally penetrate more readily than large ones, and lipophile molecules penetrate more readily than hydrophilic ones. In addition, we also see that polar molecules such as strong electrolytes penetrate the membrane very slowly. This fact may be correlated with the electrical properties of plasma membranes shown in Table 3, and suggest that plasma membranes may function as low-dielectric barriers (insulators).

How then do hydrophilic substances (e.g. a polar substance like water) penetrate this barrier?

The physico-chemical evidence is that they do, and there are at least two explanations. On the 'unit membrane' theory it is proposed that the lipid bilayer alternates in places with proteinaceous areas giving rise to channels or polar 'pores' through the membrane, which allow movement of water.

By 'pore' is meant a highly polar region of the membrane, that is, a region of the membrane which is more aqueous than the other fatty part. It does not imply a simple tube, and unlike the pores in the nuclear membrane they are *not* visible under the electron microscope.

Another interpretation can be seen if we look at a surface view of the Lucy model:

24

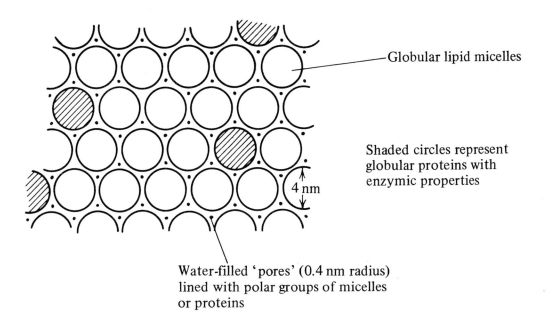

Globular lipid micelles

Shaded circles represent globular proteins with enzymic properties

4 nm

Water-filled 'pores' (0.4 nm radius) lined with polar groups of micelles or proteins

The macromolecular assembly of micelles is regarded as flexible, with individual micelles in continuous slight random movement. Again this arrangement permits us to postulate the presence of polar 'pores' between the globular micelles of lipid and protein.

Although the 'pore' concept is useful in describing some membrane properties, there is no visual evidence or other proof of their existence, except in the nuclear membrane where pores of defined dimensions are seen in electron micrographs. Nevertheless, in red blood cell plasma membranes it has been calculated that 10^5 diffusion channels, accounting for about 10^{-3} of the surface area, would describe the observed ion-permeability characteristics (Cohen *et al.* 1968).

If these diffusion channels ('pores') were uniformly distributed one per 100 nm² over the membrane surface, how far apart would the 'pores' be?

10 nm apart

24 You have seen that the membrane is selectively permeable to different substances.
How would you define permeability?

Permeability can be defined as the weight of a substance (molecule or ion) taken in or lost per unit area per unit time.

25 Table 1 showed that the plasma membrane has a higher permeability to water and gases than to other essential substances like glucose and amino acids.

Consider a spherical cell of radius r with a selectively permeable membrane of thickness d. Inside the cell there is an aqueous solution (e.g. glucose) of concentration C_1. The cell itself is completely bathed in a solution of the same solvent and solute but the glucose concentration is higher, i.e. C_2.

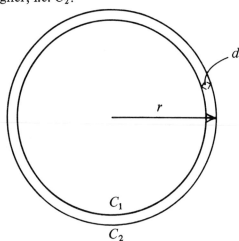

What factors determine the rate at which the solute (glucose) enters the cell?

_____

_____

(i)   The difference in the concentration of molecules on either side of the membrane (i.e. $C_2 - C_1$) ≡ net flow.
(ii) The mobility of the solute in the membrane phase, which is dependent both on the structure of the plasma membrane and on its thickness.
(iii) The surface area of the cell, $4\pi r^2$.
(iv) Temperature.

26 How do molecules manage to move from one side of the membrane to the other?

_____

_____

By diffusion. The net result is a movement from areas of high to areas of low concentration until equilibrium is reached.

27 From the situation depicted in frame 25 derive an equation for a, the rate of net inward diffusion of solute. Use the variables already given and a permeability constant, k. Assume for the purposes of the equation that diffusion is directly dependent on temperature.

_____

_____

$$a = k\,(C_2 - C_1)\,.\,4\pi r^2\,.T$$

---

28   In other words (as you correctly deduced) the rate of diffusion of a
substance is proportional to its concentration difference across the
membrane at temperature T. It is also governed by the permeability
constant of the substance (i.e. its mobility in aqueous solution and the
ease with which it can penetrate the membrane). However, from the
point of view of effective diffusion of substances to areas within the
cell, the *surface area : volume ratio* is of great significance, since
diffusion in aqueous solution, even under ideal conditions, is only
effective over a comparatively small distance. Consequently, this is one
of the key factors determining the maximum size attained by individual
cells.

Consider the following hypothetical case of three cube-shaped cells;
the first has linear dimensions of 1 μm, the second 10 μm, and the
third 100 μm.

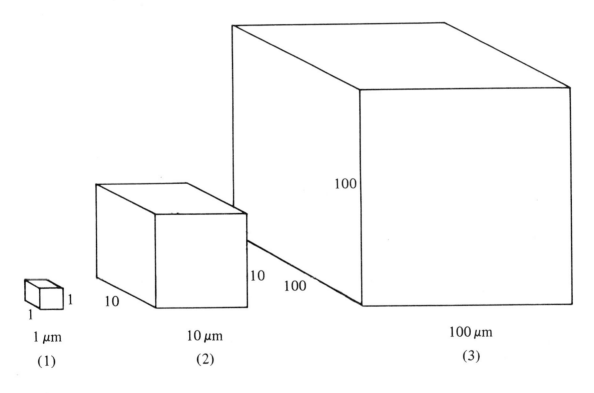

Calculate the surface area to volume ratio in each case.

In (1)
   Area = 6 μm^2
   Volume = 1 μm^3
   Ratio = <u>6 : 1</u>
In (2)
   Area = 600 μm^2
   Volume = 1000 μm^3
   Ratio = 600 : 1000 or <u>6 : 10</u>

In (3)

   Area = 60 000 μm^2
   Volume = 1 000 000 μm^3
   Ratio = $6 \times 10^4 : 10^6$ or 6 : 100

Immediately you begin to see that by reducing the dimensions of a cell, the surface area is reduced by the square root, but the volume is reduced by the cubic root.

---

29 Table 1 showed that the plasma membrane is virtually impermeable to sucrose molecules. So if a cell is placed in a strong aqueous solution of sugar, what kinds of molecules would you expect to move through the membrane and in what directions? Why?

Water molecules will move through the membrane in an outward direction because the concentration of water in a strong sucrose solution is appreciably less than that in the cytoplasm.

---

30 This special kind of diffusion in which there is net movement of solvent molecules is called *OSMOSIS*. How would you expect osmosis to affect the appearance of the cell in the case described above?

The cell would shrink through loss of water.

---

31 If a strong aqueous solution of sugar was enclosed within a membrane selectively permeable to water, and then the membrane in turn is placed in water, what will happen?

There will be a net flow of water from outside to inside and the membrane will expand.

28

32 Up to this point we have considered the plasma membrane as a passive, selectively permeable barrier which allows certain substances to diffuse across it. We shall now consider evidence where mechanisms other than diffusion and osmosis are in operation, i.e. mechanisms which involve more active chemical participation by the membrane.

It is known, for example, that after feeding, soluble foods, salts and water are absorbed into the bloodstream in the region of the small intestine. Since the network of fine blood vessels in the intestine is sufficiently elaborate to ensure that all cells lining the inside are within a few micrometres of a blood vessel, it is reasonable to suppose that soluble foods pass from the lumen of the gut through these lining cells into the blood.

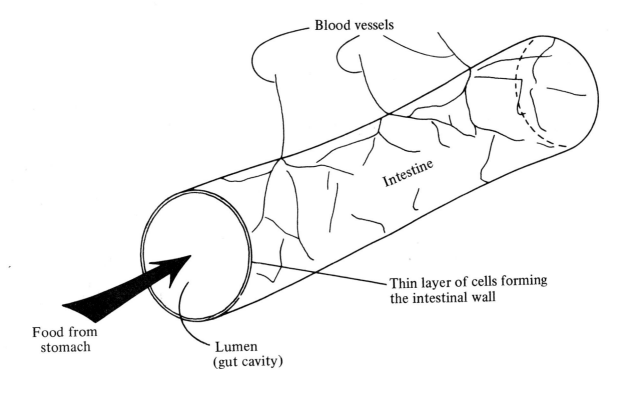

(a) How would you expect the concentration of soluble foods and salts to vary between the lumen of the gut and the blood supplying the intestinal lining one hour after feeding?
(b) What consequences would you expect?

_____

_____

(a) A relatively high concentration in the lumen of the gut and a relatively low concentration in the blood supply.
(b) Diffusion of molecules of soluble food and salt from the gut lumen to the blood vessels.

_____

33 Under certain experimental conditions it is possible to tie off sacs of the small intestine and investigate their properties in isolation from the rest of the digestive tract. The concentration of sodium chloride (NaCl) in the blood plasma is about 1%. So in one experiment three sacs were tied off (A, B and C) and filled with NaCl solution of the following concentrations

     A   0.5%
     B   1.0%
     C   1.5%

If you assume that only passive transport, i.e. diffusion and osmosis is involved, what movements of salt molecules and water molecules would you expect for each sac?

A.   Net gain of NaCl by diffusion into the solution contained in the sac; net loss of H_2O by osmosis out of the sac
B.   No change
C.   Net loss of NaCl by diffusion out of the solution in the sac; net gain of H_2O by osmosis into the solution in the sac

34 In fact there was a net loss of both NaCl and H_2O from all three sacs.

Table 5

Sac	Amount put into intestinal sac		Amount absorbed by cells of intestine	
	H_2O (ml)	NaCl (g)	H_2O (ml)	NaCl (g)
A	100	0.5	75	0.3
B	100	1.0	50	0.6
C	100	1.5	25	1.0

What conclusions can you draw?

Some transport mechanism other than diffusion and osmosis must be operating.

35 Do these data indicate that diffusion and osmosis are partly involved?

They suggest that they are, because it is possible that the 'other transport mechanism' could be responsible for the transfer of similar amounts of water and sodium chloride in all three sacs with diffusion and osmosis accounting for the different net amounts transferred from each sac.

    However, other interpretations of the data are also possible; e.g. the other mechanism could be concentration-dependent.

36   In another experiment an intestinal sac was filled with glucose solution containing sodium sulphate (the importance of adding a sodium salt will be apparent later). The normal concentration of glucose in the blood is 180 mg per 100 ml of solution (or 10 mM glucose) and it does not significantly diverge from this level. The initial glucose concentration in the sac, however, was 250 mg glucose per 100 ml solution. Samples were withdrawn at intervals and analysed for glucose.
The results are presented graphically.

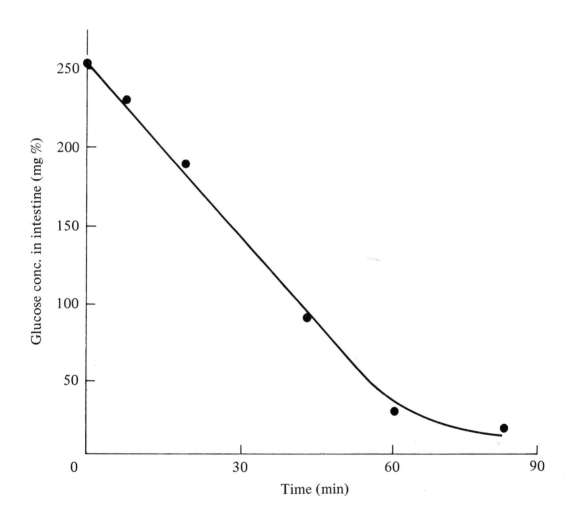

To what extent can these results be accounted for by diffusion alone?

_____

_____

Diffusion alone could only account for the movement of glucose in the first 25 minutes of the experiment.
Note. If diffusion was an important factor, there would be some deviation after 25 min from the straight line graph shown here.

37   We now have evidence that both NaCl and glucose can be absorbed through the intestinal wall against their concentration gradient. This cannot be explained by the processes of diffusion and osmosis in which the cells of the intestinal lining assume a purely *passive* role. What alternative mechanism can you suggest?

_____

_____

There must be some *active* process at the surface of or within the membranes of these cells which 'pulls' or 'pumps' these molecules into the bloodstream.

---

38   Though this *'active transport'* process is presumably chemical rather than physical it can be considered as a 'pumping' process, in that moving molecules against a concentration gradient is analogous to pumping water up a hill (gradient).

   Both processes do work and hence require a source of e_____.

_____

_____

energy

---

39   Cells obtain most of their energy from the breakdown of sugars, a process which usually involves the consumption of molecular oxygen. If the 'active transport' mechanism is indeed connected with sugar metabolism and with oxygen consumption, we should expect to find that:

   (*a*)   glucose was not only transported against a concentration gradient but also _____; and

   (*b*)   oxygen _____

_____

_____

   (*a*)   used up in order to provide energy for the 'active transport' pump
   (*b*)   was necessary for 'active transport'

---

40   Both these hypotheses were tested by some experiments by Wilson & Wiseman on *everted* sacs of the small intestine. To evert a sac one turns it inside out (rather like a glove) so that the outside (*serosal* side) becomes the inside and the inside (*mucosal* side) becomes the outside.

   The process is shown on the opposite page.

(1) *In situ*

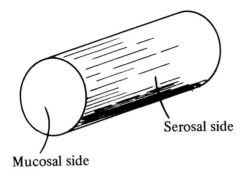

Serosal side

Mucosal side

(2) Eversion

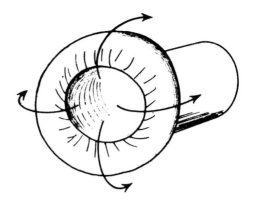

(3) After eversion

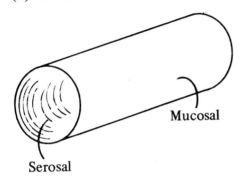

Mucosal

Serosal

(4) Ligatured ends forming
an everted sac

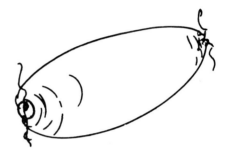

What is the experimental advantage of using an everted sac?

_____

_____

In the earlier experiments with glucose we could only measure glucose
loss from the inside. Since glucose transported through the wall of the
intestine was absorbed into the blood it was impossible to measure how
much of the glucose lost from the mucosal side was transported and
how much was metabolized 'en route'. With an everted sac the blood
supply is disconnected and all the glucose transported across the
intestinal wall remains within the sac where its concentration can be
measured.

_____

41   The experiments were conducted on adjacent pieces of hamster
intestine under two conditions: (A) aerobic (oxygen present) and (B)
anaerobic (no oxygen present). Each sac was filled with about 0.5 ml
of glucose solution (containing sodium sulphate) and placed in a bath
containing about 5 ml of the same solution. After 1 hour the sacs were
removed and the glucose concentrations on both sides measured.

CELL MEMBRANES

Table 6

Everted sacs of hamster intestine (conditions)	Oxygen uptake (QO_2)	Concentration of glucose in mg/100 ml		
		Initial conc. both sides	Final conc. serosal	Final conc. mucosal
A (aerobic)	20.5	342	782	168
B (anaerobic)	0	342	268	268

What do these results suggest about the effect of oxygen on 'active transport'?

_____

_____

They suggest that oxygen is necessary for active transport.

42 With reference to Table 6, what hypothesis can you suggest to account for the decrease in glucose concentration under anaerobic conditions?

_____

_____

That it is being used by the cells of the small intestine by a mechanism which (a) does not involve molecular oxygen and (b) does not lead to the provision of energy for active transport.

43 This complicates the evidence for our other hypothesis in frame 39, that glucose can serve as an energy source for active transport. However, the results for the quantity of glucose transferred and the quantity of glucose metabolized (here measured in μmoles) were as follows:

Table 7

Everted sacs of hamster intestine (conditions)	Net gain of glucose on serosal side (μmol)	Glucose metabolized (μmol)
A (aerobic)	+ 394	300
B (anaerobic)	− 170	365

(a) Sacs A and B could differ in surface area by as much as 50%. So is it meaningful to compare the 300 with the 365 μmol?

(b) Can we deduce, however, that under anaerobic conditions 170 μmol of glucose were metabolized on the serosal side? Explain your answer.

_____

_____

34

(a) No

(b) No. This might be true; but it is equally possible that the loss on the serosal side was caused by transfer to the mucosal side in order to maintain the same concentration on both sides.

---

44 What conditions would make such a transfer necessary?

A faster rate of glucose metabolism on the mucosal side

---

45 Tables 6 and 7 show that under anaerobic conditions glucose is metabolized by a pathway that does not involve oxygen; and which does not lead to active transport. Under aerobic conditions there are a number of possibilities. Which of the following three would you consider to be the most fruitful hypothesis? Why? (You do not yet have sufficient evidence to prove any of them, but your choice could determine the kind of evidence you would look for.)

(1) Glucose metabolism is identical under both aerobic and anaerobic conditions, i.e. it does not require oxygen and does not promote active transport.

(2) Glucose metabolism is entirely different under aerobic conditions, requires oxygen and promotes active transport.

(3) The anaerobic mechanism still operates under aerobic conditions, but it is in competition with an additional aerobic mechanism which requires oxygen and promotes active transport.

The similar quantities of glucose metabolized would appear to support (1), but this is not very strong evidence. If (1) is true then we still have to find an energy source for active transport. Oxygen by itself is only an energy source if there is something to oxidize.

(2) could also be true, but why should the anaerobic mechanism cease to operate? The anaerobic mechanism could be inhibited by oxygen, but why introduce this extra complication unless we have to.

(3) seems to be the best bet as this stage.

---

46 In fact Wilson & Wiseman were able to show that virtually all of the glucose metabolized under anaerobic conditions was converted to lactic acid (1 molecule of glucose giving 2 molecules of lactic acid).

How much lactic acid would you expect to have found under aerobic conditions if: (a) hypothesis (1) was true
                    (b) hypothesis (2) was true
                    (c) hypothesis (3) was true?

Note. You will need to refer back to Table 7.

(*a*)   about 600 μmol
(*b*)   none at all
(*c*)   some value intermediate between 0 and 600 μmol.

---

47   The results confirmed (3) and the relative importance of the two pathways was shown to vary from one part of the intestine to another. Wilson & Wiseman also showed that under aerobic conditions lactic acid tended to accumulate on the serosal side. What does this suggest?

That lactic acid, as well as glucose, can be actively transported against a concentration gradient.

---

48   We still have not explained why we added sodium sulphate to our glucose solution in all these active transport experiments. Can you think of two possible reasons?

The most obvious include the following:
(1)   that ionic strength is important;
(2)   that sodium ions in particular are needed;
(3)   that sulphate ions in particular are needed.

---

49   How would you test these hypotheses?

By observing whether active transport of glucose took place in the absence of sodium sulphate, or when sodium sulphate was replaced by another sodium salt or another sulphate.

---

50 Now look at the data below.

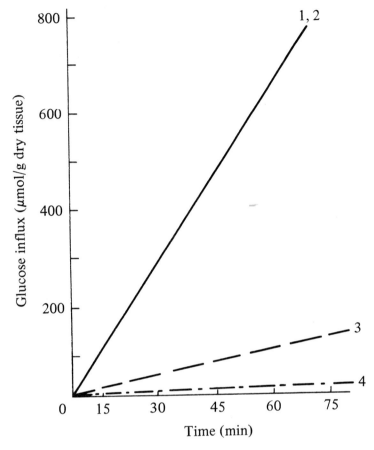

1. With Na_2SO_4 in the glucose solution
2. With NaCl in the glucose solution
3. With Li_2SO_4 in the glucose solution
4. No salt present

What do these results demonstrate?

_____

_____

That glucose uptake is dependent on the presence of a salt, and that sodium salts are much more effective than lithium salts.

51 It might be supposed that sodium-dependent glucose uptake is a peculiar feature of gut cells. When Schultz *et al.* (1966) investigated the uptake of alanine (an amino acid) by intestinal cells they obtained the results overpage:

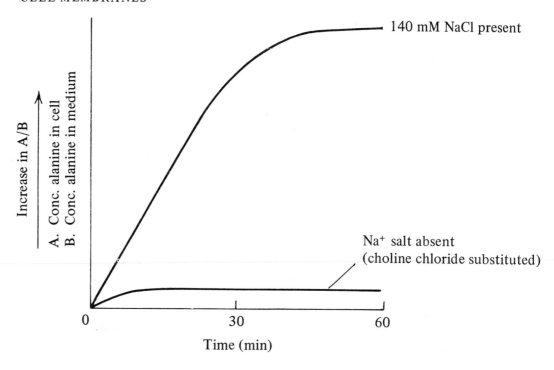

Explain these data.

_____

_____

The uptake of alanine is also sodium-dependent.
Choline chloride is substituted to maintain osmotic comparability.

---

52 In fact, it has been shown that the *active* transport of all amino acids investigated is Na^+-dependent. Which amino acids would you expect to be absorbed fastest if both diffusion and Na^+-dependent active transport are involved?

_____

_____

Those with low molecular weight.
    This prediction is often true for *neutral* amino acids, so that glycine for example is faster than phenylalanine. However, the situation is not as simple as we have implied here. For example, when an amino acid carries a charge (usually a positive charge), the rate of absorption may be slower.

---

53 Another feature of Na^+-dependent amino acid uptake is that it appears to occur in many cells. Na^+-dependent glucose uptake however, is restricted to epithelial cells (e.g. the cells lining the gut).
How then is glucose likely to be absorbed by other cells?

_____

_____

By diffusion.

Note. In some cells (e.g. red blood cells) glucose enters the cell by 'facilitated diffusion'. This is a process which does not involve 'active transport' (i.e. an energy source), but one in which the rate of transfer of glucose is greater than can be accounted for by the forces of simple diffusion based on kinetic studies.

It is possible that a 'carrier molecule' on the outer surface of the membrane may combine with glucose, move with it across the membrane and then release the glucose on the inner face of the membrane.

54   Up to this point we have only considered the active transport of glucose and amino acids across the cell of the small intestine.

Let us take as another example, a cell with which you are already familiar (Book 1) — the red blood cell or erythrocyte. Because of its easy availability for experimental work, the red blood cell has been the starting point for much intensive research into membrane transport. Analysis of the Na^+ and K^+ content of fresh human red blood cells shows that they contain about 25 mM Na^+ and 140 mM K^+ inside. The concentrations of these ions in the blood plasma (i.e. the fluid component of the blood surrounding the blood cells) are 150 mM Na^+ and 5 mM K^+.

What mechanism is responsible for maintaining these concentration gradients?

Active transport is the most likely explanation, although the concentration gradients could just conceivably arise as a result of the membrane's impermeability to both Na^+ and K^+.

55   Assuming that active transport is the correct explanation, in which direction does active transport move (*a*) Na^+ ions and (*b*) K^+ ions?

(*a*) Na^+ ions are moved out of the cell.
(*b*) K^+ ions are moved into the cell.

56   What forces are tending to diminish these actively maintained concentration gradients?

Diffusional forces. K$^+$ will tend to diffuse outwards, and Na$^+$ will tend to diffuse inwards.

---

57 In all cases dealt with so far, we have been interested in the *net flux* of substances, i.e. the extent to which the rate of forward migration exceeds the reverse movement and so produces a net transfer.
Now consider the following experiment in which dead cells and living cells were placed in concentrated (250 mM NaCl) salt solution.
The results were as follows:

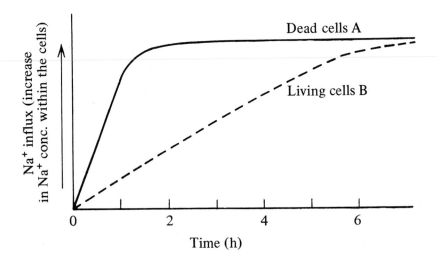

How would you explain these results?

In A, Na$^+$ diffuses into the cell until equilibrium is reached between the concentrations outside and inside the cell. There is no active transport mechanism because the cells are dead.
In B, the inward diffusion of Na$^+$ is offset by active transport which is pumping Na$^+$ out of the cells against a concentration gradient. However, at this abnormally high concentration the pump is always fighting a losing battle, and it eventually fails altogether. (If it had not failed there would be a slightly 'improved' final equilibrium.)

---

58 Now let us look at an experiment performed by Post & Jolly in 1957. Human red blood cells were stored at 2 °C under aerobic conditions in 150 mM NaCl. A metabolizable substrate (e.g. glucose) was also present. The observed differences in behaviour were as follows:

Table 8

	Na$^+$ conc. *inside* cell (mM)	K$^+$ conc. *inside* cell (mM)
A Normal (37 °C)	25	140
B After 24 h at 2 °C	110	20

(a) What has happened to the concentrations inside the cell of (i) Na⁺ ions and (ii) K⁺ ions?

(b) What hypotheses can you put forward to account for this result?

(a) (i) The Na⁺ concentration inside the cell has *risen*.

    (ii) The K⁺ concentration inside the cell has *fallen*.

(b) The active transport process (like most chemical processes) is temperature-dependent. Moreover, it is more sensitive to reductions in temperature than diffusion, which is a purely physical process.

Whereas active transport predominates at 37 °C, diffusion is predominant at 2 °C.

---

59 Is the 'pump' still operating at 2 °C, or has it been totally inactivated?

It would appear to be operating because there is still a concentration gradient; alternatively it is just possible that the cell contents have not yet reached equilibrium.

---

60 Next, Post & Jolly transferred the red blood cells to a fresh solution of 150 mM NaCl, containing a metabolizable substrate; and incubated them at 37 °C for 2 h.

No significant change occurred in the Na⁺ and K⁺ concentrations inside the cells. How does this result affect your hypothesis of frame 58?

Clearly temperature is not the only factor involved in the inactivation of the pump.

---

61 The same red blood cells were then divided into two groups and treated as follows:

Group 1: placed in 130 mM NaCl + 20 mM KCl + metabolizable substrate at 37 °C

Group 2: placed in 150 mM NaCl + metabolizable substrate at 37 °C.

The results are depicted graphically overpage:

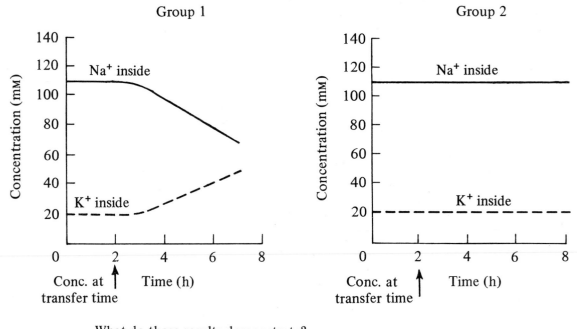

What do these results demonstrate?

_____

_____

The 'pump' will not operate unless K$^+$ is present outside the cell. The expulsion of Na$^+$ against its concentration gradient and the absorption of K$^+$ against its concentration gradient appear to be linked processes.

62 Why were the Group 2 cells placed in 150 mM NaCl rather than 130 mM NaCl?

_____

_____

Group 2 was a control group, and the purpose of a control group is to ensure that *only* the important variable is being investigated, in this case the presence or absence of K$^+$. Without the extra 20 mM NaCl there would have been an additional difference between the two groups: namely, ionic strength. Identical ionic strengths are necessary for the maintenance of osmotic compatibility.

63 We have now examined the effects of two variables in the Na$^+$/K$^+$ active transport system: namely, temperature and the presence of K$^+$ in the surrounding medium. Can you think of another variable which is likely to affect the pump and of a simple experiment to investigate its effect?
    (*Hint.* The variable has been mentioned earlier.)

_____

_____

The presence or absence of an energy source, such as glucose (metabolizable substrate).

42

Incubation of red blood cells in NaCl/KCl at 37 °C
(a) with glucose added and (b) without glucose.

64 This experiment was carried out by Whittam in 1958, when he measured the rate of K^+ influx using a radioactive isotope.

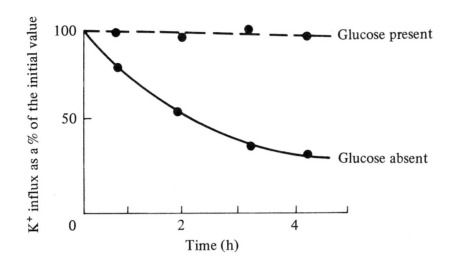

How would you interpret this result?

A metabolizable substrate is necessary for active transport. Glucose is one such substrate, but other substrates such as fructose, adenosine and inosine can serve the same purpose, and for some experiments are more suitable than glucose.

65 The retardation of K^+ influx in the lower curve in frame 64 is relatively slow, although it is known that the concentration of glucose in the cells decreases very rapidly indeed due to a combination of metabolic breakdown and diffusion outwards into the glucose-free incubating medium. Does this evidence suggest that glucose is the immediate source of energy for active transport? Explain.

No. The most reasonable explanation would seem to be that it is a product of glucose metabolism rather than glucose itself which is the *immediate* energy source for active transport.
The observation can be explained by postulating that the product cannot diffuse outwards through the membrane, and therefore decreases in concentration much more slowly than glucose itself.

66 Now the key compound in many intracellular energy transfer reactions (actually phosphate group transfer) is known to be a substance called ATP (<u>A</u>denosine <u>T</u>ri<u>P</u>hosphate). As ATP is an energy source for a number of important reactions, it was natural to see whether ATP was the *immediate* energy source for active transport. How would you test this hypothesis?

By seeing whether K^+ influx was proportional to the concentration of ATP.

---

67 Whittam's experiment gave the following results:

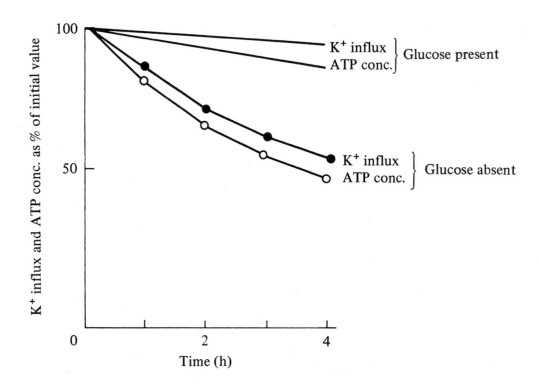

Do these results prove unequivocally that ATP is the immediate energy source for active transport?

No, there could be some intermediate whose concentration curve was exactly the same.
But it is clear that the breakdown rate of ATP is correlated with the rate of K^+ influx, and it is highly probable that ATP is the immediate energy source.

(The possibility that ATP is a product of the energy changes accompanying active transport is ruled out by our knowledge of ATPs 'position' in phosphate group transfer reactions – for a more detailed discussion of this issue you are referred to Book 8, *Metabolism and mitochondria.*)

---

68   It is clearly of interest to establish exactly where this transfer of energy from ATP to the active transport 'pump' takes place, or, to look at the problem another way, to find where the breakdown of ATP occurs.
Can you think of two possible places?

(1)   In the cytoplasm
(2)   In the membrane itself. (Remember the Lucy model in frame 20.)

---

69   A particularly elegant way of investigating the site of ATP breakdown is suggested by an unusual property of red blood cells, the ability to form 'ghost cells'.
Like other cells, red cells expand when incubated in very dilute NaCl, and eventually burst. This process is usually referred to as 'haemolysis', because it involves the loss of haemoglobin, the important compound in blood which carries oxygen and carbon dioxide.
What is the cause of haemolysis?

Osmosis. Water diffuses into the cell because the cytoplasm exerts a higher osmotic pressure than the surrounding medium. Eventually the cell becomes so large that the membrane cannot be stretched any further and gives way.

---

70   The osmotic and pH conditions for haemolysing and making 'ghost' cells are fairly critical. Briefly, the red cells are washed in isosmotic salt solution (see Glossary) and then haemolysed in a weak salt solution (40 mosmol) at pH 3.2 for 5 minutes at 0 °C. The haemolysed red blood cells consist of the plasma membrane and a little of the cytoplasmic contents.
If NaCl (usually 3 M) is added to give an isosmotic incubation medium, the membrane reconstitutes itself approximately to its original shape and regains its selective permeability. If the haemolysed cells are well washed, then these reconstituted 'ghost cells' will contain mainly the incubating medium in which they were formed.
What is the advantage in investigating active transport phenomena in reconstituted 'ghosts' of red blood cells?

If transport proceeds normally in 'ghost' cells it will provide good evidence that energy transfer from ATP to the 'pump' takes place inside the membrane rather than in the cytoplasm.

---

71   The energy for driving the 'pump' is derived from the hydrolysis of ATP (adenosine triphosphate). The reaction is known to give ADP (adenosine diphosphate) and inorganic phosphate ions:

$$ATP^{4-} \xrightarrow[(H_2O)]{} ADP^{3-} + HPO_4^{2-} + 30J/mol \qquad (\equiv \text{heat energy})$$

so ATP breakdown can be followed by observing changes in the concentration of phosphate ions.

In a series of experiments with red blood cell ghosts, into which ATP had been incorporated by incubating the cells in a haemolysing medium containing ATP, Whittam & Ager (1964) measured the initial rate of K^+ influx and the initial rate of increase in phosphate ion concentration. In each case the concentration of K^+ outside the ghost cells, i.e. in the incubating medium, was different. The results are shown below.

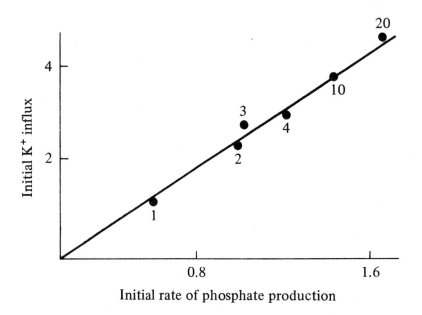

Initial rate of phosphate production

How would you interpret these results?

Very strong evidence that:
(1)   The intake of K^+ and the production of phosphate are directly proportional to each other; and
(2)   ATP breakdown is associated with the plasma membrane.

---

72   In another experiment, ATP was not present when the ghosts were made up, but was added subsequently to the incubation medium. No active transport took place and no inorganic phosphate was produced.

What did this show?

That external ATP could not be used as an energy source. Only ATP inside the cell can be used.

---

73   Earlier experiments by Whittam (1962) with ghost cells had shown that internal Na^+ concentration affects the rate of phosphate production. In each case the total initial internal ionic strength was the same, 145 mM, and the external medium was 140 mM Na^+, 5 mM K^+. The internal ATP concentration was 4 mM.

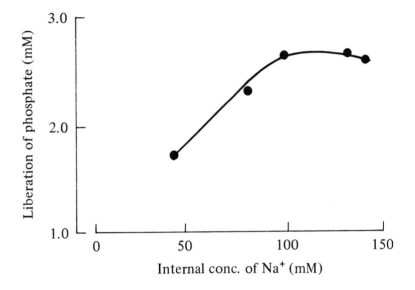

What do these results suggest about the relationship between internal Na^+ concentration and phosphate production?

The rate of phosphate liberated is dependent on internal Na^+ up to 100 mM. Extrapolation to even lower Na^+ concentrations suggests that ATP breakdown will be minimal unless some internal Na^+ is present.

---

74   In frame 61 we commented that the intake of K^+ and the expulsion of Na^+ appeared to be linked processes, and showed that external K^+ was necessary for active transport.

In frames 67 and 71 we showed that the intake of K^+ and the breakdown of ATP were linked.

We now have evidence that internal sodium is linked with ATP breakdown. Although we do not know any more about how ATP breakdown is linked with both Na^+ expulsion and K^+ intake, we do know that it takes place within the membrane and that it uses internal ATP derived from glucose, or possibly from other metabolizable substrates.

Would you say that this evidence from 'active transport' mechanisms was consistent with the 'unit membrane' model or one of the other models presented earlier, when considering the structure of the plasma membrane?

An answer is *not* given here, because there is no definitive answer, but it is a question which you might like to think about or discuss with your colleagues.

We shall return to the problem of active transport and the distribution of Na$^+$ and K$^+$ ions, when we look at nerve and muscle physiology in Book 10 of this series.

5.2. Phagocytosis and pinocytosis

In this section we shall investigate ways in which bulk solids or fluids can get into and out of certain cells.

For the next 6 frames of this book, you will require the film loop entitled *Phagocytosis*. Place this film loop into the loop projector as shown in the diagram (section 5.2.1) on the next page, and then follow the operational instructions for viewing the film.

Those of you who already understand 'phagocytosis' (a definition is given in the glossary of terms at the back of this book), or who wish to omit the film, should continue with the programmed text, starting at frame 81.

5.2.1. Instructions for using film loop projector

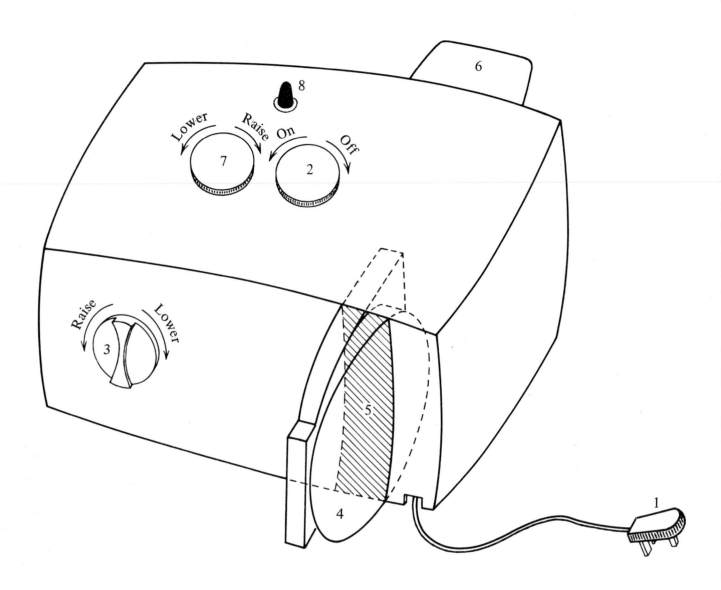

Read the following instructions through carefully before starting. Refer to
the diagram and give special attention to F, G, H and I.

A. Plug (1) into mains. Switch on the projector by turning (2).
B. Centre the projector on the screen by moving the *screen* appropriately,
    and by adjusting (3).
C. Switch off the projector by turning (2).
D. Plug the film loop (4) into the slot (5) as shown in the diagram.
E. Switch on by turning (2).
F. *Immediately* focus the image by turning (6) appropriately.
G. *If necessary* adjust the frame position by turning (7).
H. Press (8) to hold the film on a chosen frame; press again to restart.
I. Turn (2) to switch off the projector.

50

PHAGOCYTOSIS AND PINOCYTOSIS

5.2.2. Phagocytosis

75 The sequences on view in this loop show human blood cells and bacteria *in vitro*, i.e. the cells are present in a thin film of blood outside the body. The sequences are numbered(1—4) for clarity.
In setting up an experiment for cine photomicrography, what experimental precautions would need to be taken?

___

Precautions to ensure that:
(i)   body temperature is maintained (i.e. 37 °C);
(ii)   the concentrations of nutrients and oxygen do not fall below a critical level,
and (iii) the elements favouring clotting of the blood are removed.
All of these factors could affect the proper working of the cells.

76 What other physical conditions would the blood cells be subjected to in the human body which they are not subjected to here?

___

All blood cells would be constantly subjected to turbulent movements of whole blood as a result of the heart's pumping action and the expansion and contraction of blood vessels.
(This factor too could alter the behaviour of the cells and should be borne in mind when assessing the photographic evidence.)

77 In the film you are about to see, the movements of the cells have been speeded up approximately fifty times. Now observe the sequences at least twice before attempting to answer the following questions.
In each of the sequences at least three different types of cell can be seen. What are they?

___

(i) Red blood cells (erythrocytes);   (ii) white blood cells (leucocytes); and   (iii) bacterial cells.

78 How many different types of bacterial cell can you see? Describe them.

___

Two.

The large rectangular ones are called *bacilli*.

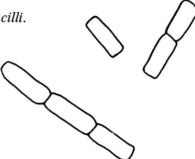
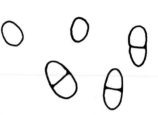

The smaller ones are called *pneumococci.*

These particular pneumococci are responsible for pneumonia.
Some of the pneumococci are paired — the result of division without separation.
In the last sequence the pneumococci have been deliberately clumped together chemically by the experimenter.

---

79 (i) What structural and behavioural features enable you to distinguish between red and white blood cells?

(ii) Can you recognize more than one kind of white blood cell?

(i) *Red blood cells*
   (i) Usually round (although they are flexible and can change shape)
   (ii) Do not show intrinsic movements
   (iii) Appear to be homogeneous in structure
   (iv) Do not have a nucleus
   (They are in fact bioconcave when viewed side-on a feature which cannot be seen here, but implied by the lighter circular area in the centre of these cells.)

   White blood cells
   (i) Shape very variable and constantly changing
   (ii) Possess intrinsic movement and are extremely active (contact must be made with the glass plate for movement to occur)
   (iii) Cytoplasmic inclusions heterogeneous (many granular inclusions)
   (iv) Possess a large nucleus

(ii) Yes. In fact, during the sequences of this film loop three different types of white blood cell are encountered: (*a*) neutrophils (sequences 1 and 4), i.e. those which stain with neutral dyes and possess a multilobed nucleus; (*b*) eosinophils (sequence 2), i.e. those staining with the dye eosin; they show a densely granular cytoplasm and possess a multilobed nucleus; and (*c*) monocytes (sequence 3), these possess a large, single unlobed nucleus.

---

80   However, the feature of prime interest in this film loop is *Phagocytosis* or 'cell-eating' (Greek *phagein* meaning 'to eat' and *kytos* meaning 'cell').
Which of the following most accurately describes the cell movements involved in phagocytosis?
(*a*)   The white cells flow around the bacteria thus engulfing them within the cytoplasm.
(*b*)   White cells appear to be attracted towards the group of bacteria and make contact with them. Phagocytosis is brought about by white cells flowing around the bacteria, eventually engulfing them in cytoplasmic vesicles.
(*c*)   The plasma membrane breaks down as white cells come into contact with bacteria. As a result the bacteria become engulfed in the cytoplasm.

(*b*)

Now turn *off* the film loop

81   What would appear to be the purpose of phagocytosis in white blood cells?

(i)   Protection against bacterial infection
(ii)   Removal of dead cells or cell debris

---

82   What hypothesis can you suggest to account for the movement of the phagocytes towards bacteria?

Bacteria release certain chemical substances which attract the phagocytes.

---

83   Can you think of an experiment which would test this hypothesis?

(Your answer)

---

84   There are a number of experiments which you might have thought of.
      One such experiment, modified after Francis (1965), investigated the
      response of certain amoeboid-like cells (see Glossary) to a central
      source of living material, which was thought to be releasing a chemical
      attractant. The experiment was conducted in a flow chamber, and the
      amoeboid cells placed into the chamber at random (situation A below).
      In a similar chamber a centrally located glass disc of similar size was set
      up as control. Again the amoeboid cells were placed in the chamber at
      random (situation B below).

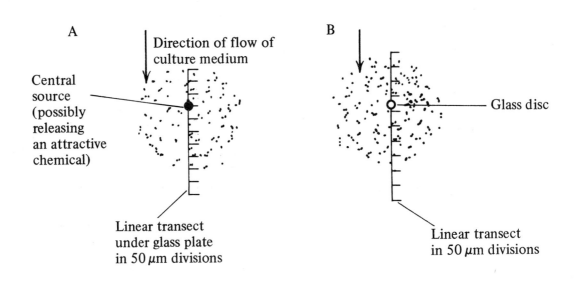

Distribution of cells after 30 min

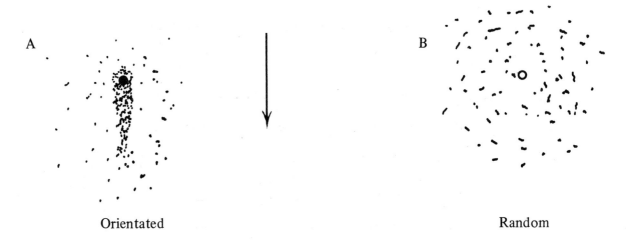

A   B

Orientated   Random

When cell counts were made along the transect line at various distances from the central source the following results were obtained:

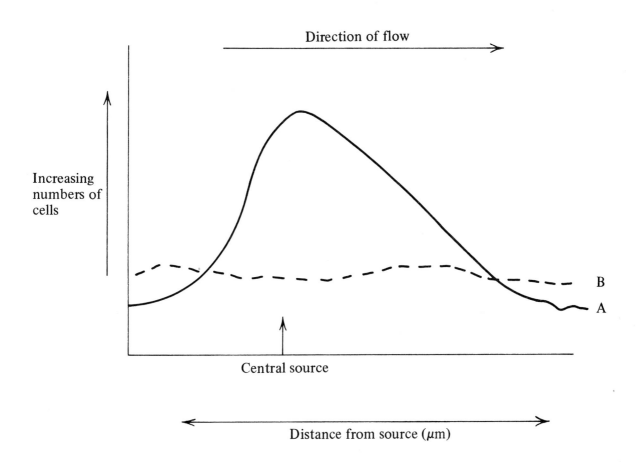

Do these results support the hypothesis? Why?

_____

_____

Yes.
If the source in situation A was releasing an attractive chemical substance, then we would predict a concentration of this chemical downstream.
The cells have orientated themselves along this gradient with the greatest number of cells nearest the source. The further away from the source, the less effective is the chemical attractant, so that there is a threshold distance below which there is no chemical influence and the distribution of cells is random.
(*Note.* This experiment precludes *in this case* another possibility, that attraction is caused by the influence of an electrical field — can you see why?)

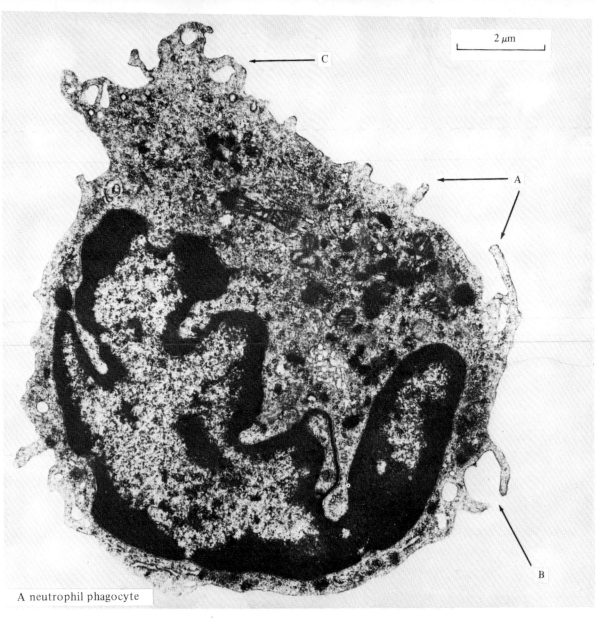

2 μm

C

A

B

A neutrophil phagocyte

85 Examine the electron micrograph of a neutrophil phagocyte, and note the large, lobed nucleus.

Now look at the plasma membrane and accompanying cytoplasm at the areas marked A, B and C. Each letter shows a stage in phagocytosis.

(i) How does this picture help you to understand what is going on in the process of phagocytosis?

(ii) Describe the sequence of events. (*Hint.* Remember that you have 2-D evidence in front of you, but that you have a 3-D problem.)

(i) It shows more clearly what is happening at the surface of the plasma membrane.

(ii) A. An early stage in phagocytosis in which projections of the cytoplasm (called pseudopodia) are extending outwardly.

B. The next stage in which the pseudopodia are forming a cup-like vesicle.

C. The final stage where the pseudopodia have fused giving rise to a spherical vesicle.

Note. At a stage between B and C there is a breakdown or change in the plasma membrane where fusion of the pseudopodia occur.

---

86 In the micrograph, the vacuoles at C appear empty. What explanations can you give for this fact?

(i)   The plane of section does not cut through the engulfed particle.
(ii) Possibly there is no particle.

87 One of the interesting problems in phagocytosis which concerned biologists for a number of years was the nature of the physical or chemical stimulus that triggers the sequence of events which culminates in the formation of a phagocytic vacuole. The table below summarizes the experimental results obtained by a number of workers in relation to the problem posed above:

Type of stimulus	*Phagocytic response*
1. Physical contact with a number of substances	Sometimes positive, sometimes negative
2. Chemical stimulation by substances bearing a positive charge (i.e. cationic substances)	Always positive
3. Chemical stimulation by substances bearing a negative charge (i.e. anionic substances)	Always negative
4. Chemical stimulation by substances bearing no charge	Always negative
5. Chemical stimulation by neutral substances complexed with cationically charged particles	Always positive

What conclusions do you draw from these data?

(i)   That physical contact in itself does not necessarily produce phagocytic activity.
(ii) In all cases examined, organisms or substances engulfed by phagocytic cells are cationic in nature or need to be made so in order to evoke a phagocytic response.
(iii) Possibility (ii) mentioned in frame 86 is ruled out.

88 You saw in the film loop bacteria of the pneumococcus type being phagocytozed by white blood cells. Yet it is known that pneumococci are extremely resistant to phagocytosis. If this is so what conclusion do you draw about the physico-chemical nature of these bacteria?

They are *not* cationic in nature

---

89 How could they be rendered susceptible to phagocytosis?

By complexing them with a cationic substance

---

90 The secret of immunity to bacterial infection lies in the ability of the human body to produce *antibodies*. These are specific proteins produced in the body which inactivate specific bacteria or invading foreign materials. The latter are referred to as *antigens*. Therefore_____ are produced by the body to counteract the activity of_____.

antibodies; antigens

---

91 Either the body produces antibodies in the form of antitoxins to counteract the toxins produced by certain bacteria, or it produces antibodies which alter the surface membrane properties of bacteria. *Opsonins* are antibodies produced in the bloodstream which alter the surface membrane properties of antigenic bacteria, rendering the latter more susceptible to phagocytosis.
What deductions do you make about the physico-chemical properties of *opsonins*?

(i)   They are proteinaceous antibodies.
(ii) They are cationic substances.
Note. In the days before antibiotics (e.g. penicillin) were used for the treatment of pneumonia, the body's production of opsonins marked the turning point in the progress of the disease.

---

92 Now turn *on* the film loop again, and look particularly carefully at the last sequence (4). At the beginning of this sequence you see a clumped mass of pneumococci amongst a number of red cells. What happens to this clump of bacteria in the course of the film sequence?

It is subjected to massed phagocytic attack.

---

93   Describe the behaviour of the white blood cells (phagocytes) in this
       particular sequence.

_____

_____

       (i)   White cells migrate by chemical attraction into the area occupied
              by the clumped pneumococci, pushing their way through groups
              of surrounding red cells;
       (ii)   as more white cells congregate at the site of infection, they
              gradually encircle the clump of bacteria;
       (iii) finally they begin to break down the clump of bacteria by surface
              erosion and phagocytosis.

_____

94   It was stated in frame 78 that the pneumococci had been deliberately
       clumped together by chemical means. Deduce what type of chemical
       substance has been used in this instance.

_____

_____

       An opsonin or related cationic substance.
       Opsonized bacteria tend to adhere together. The total effect is to
       render them more vulnerable to phagocytic attack.
       (One explanation of the apparent paradox which exists between
       positively charged substances repelling each other and opsonized
       bacteria adhering together might be that cross-linking occurs between
       the more negatively charged bacterial cells and the positively charged
       opsonin molecules.)

_____

95   Before leaving this film loop you may have noticed that in certain
       sequences (e.g. sequence 3) some red blood cells were curiously dis-
       torted. Bearing in mind the conditions of the experiment, how would
       you account for their distorted shape?

_____

_____

       It is difficult to explain, but one possibility is that certain parts of
       these red blood cells are adhering to the substratum (glass plate in this
       case), whereas other parts are not and this has caused the distortion in
       shape.

_____

96   There are several questions which are still unanswered. How is move-
       ment brought about? What changes occur in the cell membrane? What
       happens to the engulfed body once entrapped in a cytoplasmic vesicle?
       As yet we have no absolute answers to the first two, but we do know
       more about the third and have suggested a simple experiment which you
       might like to carry out. It is outlined in the Appendix, Experiment 1 at
       the back of the book.

_____

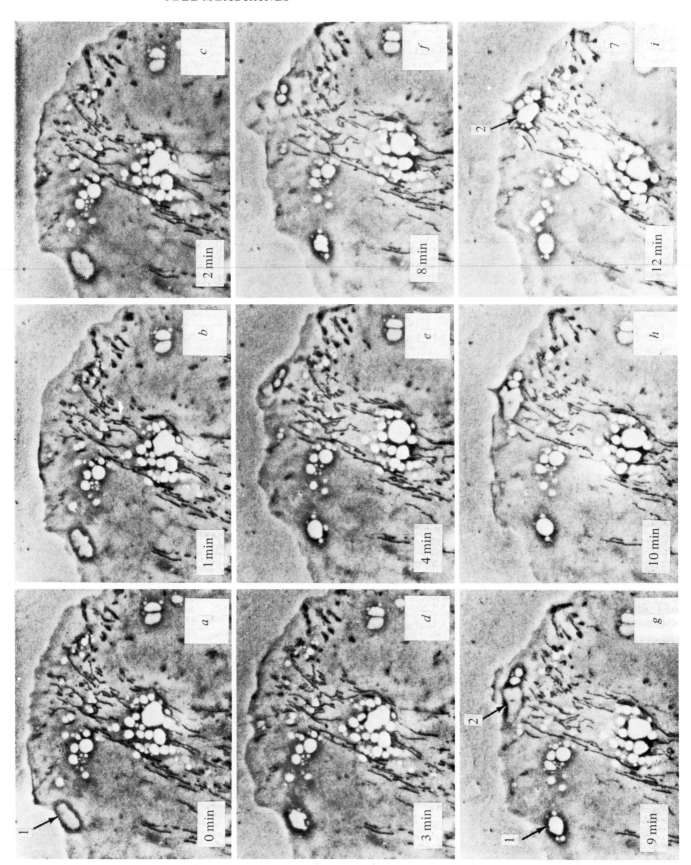

Time-lapse sequence (phase contrast) of a human sarcoma cell undergoing pinocytosis. The time between exposures is given; the magnification is ×2500. Reproduced with permission from *Cinemicrography in Cell Biology*, edited by G.G. Rose, pp. 279–312. Academic Press (1963).

5.2.3. Pinocytosis

97   The time-lapse sequence of light micrographs on the opposite page
      shows a cell in the process of 'cell drinking' or *pinocytosis** (Greek
      pinein = 'to drink'). Here a bubble-like structure (arrowed) appears to
      arise at the surface membrane of the cell and then moves into the
      cytoplasm.
          Pinocytosis is a process whereby solutes are transported into the cell
      in bulk as a result of changes at the membrane surface. Areas where
      pinocytotic vesicles form are metabolically very active, as can be seen
      by the large number of thread-like mitochondria present in these
      photographs. What possible advantage does this method offer over the
      normal process of diffusion?

      It might allow large molecules (e.g. proteins) to enter the cell, which
      would otherwise find it impossible by diffusion.

98   How could this idea be tested?

      By selecting a specific substance which is itself strongly coloured or
      stains in a discrete way with a dye, and then seeing whether it is taken
      up by a cell in the process of pinocytosis.
      (In fact, a number of experiments of this kind have been conducted.)

99   Pinocytosis has been observed in a variety of cells, although not all cells.
      It seems that pinocytosis can take place in at least three different ways:
      (1)   elevation of a ruffling membrane to entrap a relatively large
            droplet of fluid;
      (2)   the inpocketing of the plasma membrane surface to form vesicles
            of submicroscopic size; and
      (3)   semipermanent tubular invaginations of the cell membrane to
            form long channels.
      Examine the three micrographs on next page; A, B & C.

*Those of you who have studied Book 3 in this Course may have already seen pinocytosis on film.

CELL MEMBRANES

A. Pinocytosis in the leading pseudopodia of *Amoeba* (a single-cell organism) as seen under phase contrast. Magnification approx. ×2000.

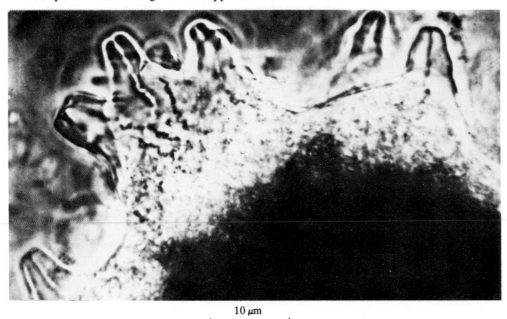

10 μm

B. Two adjacent cells forming a capillary blood vessel.

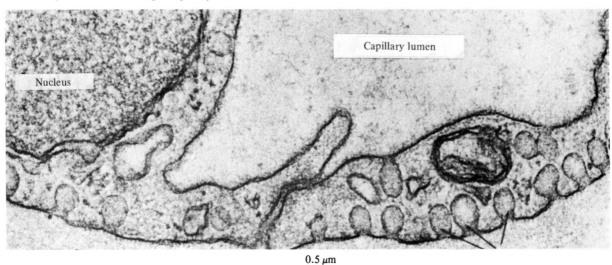

Capillary lumen

Nucleus

0.5 μm

C. An area from a similar cell to B.

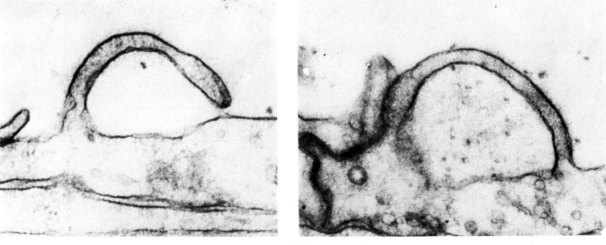

0.5 μm

Which of the three descriptions on p. 61 best matched the evidence in (i) A, (ii) B and (iii) C?

3 and A
2 and B
1 and C

100 In frame 97, it was suggested that pinocytosis might be a means by which microscopic particles or large molecules get into cells.
Examine the micrographs below showing the formation of small pinocytotic vesicles.

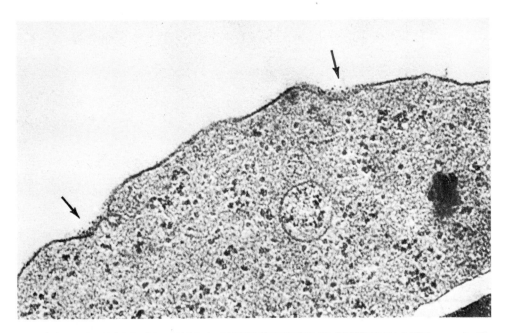

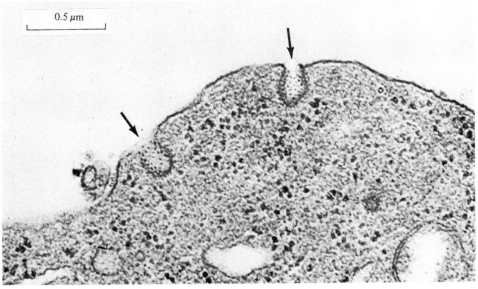

0.5 μm

Does the evidence presented in this picture support the above suggestion? Why?

_____

_____

Yes; small particles are clearly seen to be enclosed within the pinocytic vesicles (these electron-dense particles are in fact methyl ferritin, a cationic substance).

_____

101 What feature is there in common between pinocytosis and phagocytosis with respect to the nature of the stimulus eliciting these phenomena? (*Hint.* See the comment in answer to frame 100 above.)

_____

_____

In both cases, the effective agents or substances are cationic.

_____

102 Two other pieces of experimental evidence are important:
   (1) In 1933 Baldridge & Gerard noticed that when phagocytic leucocytes were presented with particulate material, there was an immediate and dramatic increase in oxygen uptake.
   (2) In a series of investigations in 1965 Zimmerman & Rustad found that metabolic inhibitors suppressed pinocytosis.
   What conclusions do you draw from these two experimental observations?

_____

_____

Both phagocytosis and pinocytosis are energy-dependent processes.
In a sense, therefore, they both represent a form of 'active transport'.

_____

5.3. Concluding remarks

In this book we have been interested primarily in the mechanisms whereby things get into and out of cells.

Firstly, as Overton originally showed, passive diffusion of substances in aqueous solution across membranes is, in part, directly related to their molecular weight as shown quantitatively from the table of diffusion coefficients at 25 °C (Table 9 below).

Table 9

Substance	Molecular weight	Diffusion coefficient ($cm^2/s \times 10^6$)
O_2 gas	32	19.8
Acetyl choline (chemical transmitter between nerve cells)	182	5.6
Sucrose (disaccharide sugar)	342	2.4
Serum albumin (protein in blood serum)	69×10^3	0.7

Nevertheless, polar molecules (e.g. strong electrolytes) show low permeability, which is not so much related to their molecular weight, but rather to the charge that they carry and their state of hydration. For example, plasma membranes have a very low permeability to Na^+ ions, but a somewhat higher permeability to K^+ ions, despite the fact that potassium has a higher atomic number (39) than sodium (23). This as we shall see in Book 10 is due to the larger 'shell' of water molecules surrounding the sodium ion, which is more hydrated than the K^+ ion.

Secondly, we have seen that cell membranes will actively transport both Na^+ and K^+ ions (as well as other molecules) under certain conditions. These active transport mechanisms require a fuel supply to maintain them. The evidence that an active metabolic process is involved can be seen, (a) by comparing the maintenance of concentration differences of ions or molecules across the membranes of living cells with those of dead cells; (b) by observing the effect on the active process when a substrate or ATP is added or not, and (c) by using metabolic inhibitors specific to the 'ATPase pump' such as the glycoside, oubain.

Thirdly, bulk solids or fluids may enter certain cells by special active processes such as phagocytosis and pinocytosis. It should be stated in these concluding remarks that 'reverse phagocytosis' (egestion) of unwanted materials does occur in some animal cells, and some cells are capable of 'reverse pinocytosis' under certain conditions. Sometimes reverse pinocytosis represents part of the mechanism whereby fluids and large molecules are carried across a cell. In other cases it may be connected with the problem of *osmoregulation*, i.e. the ability to maintain a relatively constant osmotic environment within the organism, despite osmotic differences in the external environment. A simple illustration of this principle is given in Experiment 2 in the Appendix at the back of this book. Then in Experiment 3, you are also encouraged to carry out a simple behavioural experiment with living *Amoeba*, a single-celled or acellular animal which is (1) chemotactic, (i.e. it will move towards a chemical attractant); (ii) phagocytic, and (iii) also exhibits reverse phagocytosis or egestion.

Fourthly, it should be mentioned that (i) some aspects of cell movement are discussed in Book 3 of this series, and (ii) the involvement of receptor sites on or in plasma membranes is discussed with particular reference to hormones in Book 11.

Finally, a brief word should be said about plasma membranes and antigens. You have already seen that antigens are specific substances which, if foreign to an organism, will give rise to antibody production by that organism. However, many animals produce their own antigens, presumably as part of their individual identification. But, an animal does not produce antibodies against its own antigens. In other words, it recognizes 'self'. Antigens are usually proteins associated with cell membranes; they might be enzymes, or parts of enzymes, hormone receptors, structural components, or combinations of these or other substances. Possibly the best examples of cell antigens are the blood group antigens belonging to the ABO system, originally identified by Landsteiner in 1900. Human erythrocytes may be divided into four groups depending on whether antigen A is present, antigen B is present, both antigens A and B are present, or antigens A and B are absent. The blood group antigens are genetically inherited. Knowledge of these differences is critical in blood transfusions, because if blood type A, or B, or AB, is given to a recipient of blood type O (i.e. A and B are absent), the recipient will produce antibodies against either or both antigens, with disasterous consequences.

Table 10 shows compatibility and incompatibility of the blood type antigens on transfusion.

Table 10

		Donor blood type			
		A	B	AB	O
Recipient blood type	A	√	X	X	√
	B	X	√	X	√
	AB	√	√	√	√
	O	X	X	X	√

√ = compatible
X = incompatible

Other types of blood group antigens have subsequently been discovered, noteably the presence or absence of Rhesus antigen (Rh^+) or (Rh^-).
Those of you who wish to read more about antigens, antibodies and membranes, are recommended to consult the Further Reading list in section 5.8.

5.4. Glossary of terms used

Active transport   an energy-requiring process which permits the transfer of substances against a concentration gradient, or at a faster rate along a gradient than can be accounted for by simple diffusion.

Aerobic respiration   cellular respiration which is ultimately dependent on molecular oxygen.

Amoeboid-like cells   cells resembling the shape and in many respects the behaviour of the single-cell organism *Amoeba* (a protozoan); characteristically the cell is irregular in shape and is capable of extending portions of the cytoplasm as pseudopodia ('false-feet').

Anaerobic respiration   cellular respiration not requiring the presence of free oxygen.

Antibody   a protein produced by an animal in response to a specific antigen.

Antigen   a substance (usually a protein or polysaccharide) which when introduced into the body of a foreign animal, stimulates that animal to produce an antibody.

Ciliates   single-celled (or acellular) organisms belonging to the phylum Protozoa moving by means of cilia (hair-like processes) which cover the cell. These Protozoans are bounded by a fairly rigid pellicle which gives them a definite shape.

Diffusion   migration of molecules or ions, as a result of their own random movements from a region of higher concentration to a region of lower concentration.

Electrolyte   a substance which ionizes in aqueous solution and whose behaviour is distinguished by its ability to conduct an electric current.

Facilitated diffusion   transfer of substances across biological membranes at faster rates than can be accounted for by simple diffusion, in which carrier or porter molecules in the membrane are thought to be involved. Evidence for these 'faster rates' comes from kinetic studies.

Freeze-etch preparations   a method of preparing biological materials for the electron microscope. The biological material (in 20% glycerol) is rapidly frozen under vacuum (to -100 °C). The specimen is then cleaved with a knife or razor blade along lines of weakness (usually membrane surfaces); the specimen is then etched by allowing water vapour to sublime from its surface; a celloidin replica made of the cleaved surfaces and the replica floated off. The replica is then coated (shadowed) with a heavy metal (e.g. platinum and carbon). The technique of freeze-fracturing is slightly simpler, i.e. the etching is omitted, but produces similar results.

Hydrophilic   literally 'water-loving' (cf. hydrophobic = water-hating.)

Intestine (Small intestine + large intestine)   part of the alimentary canal following the stomach in mammals. The small intestine comprising the duodenum, jejunum and ileum is concerned with the digestion and absorption of foods. The large intestine (smaller in length, but with a wider lumen) is mainly concerned with the absorption of water.

Isosmotic   literally 'the same osmotic pressure'; i.e. a solution is said to be isosmotic with a cell when both have the same osmolarity or exert the same osmotic pressure. Compare hyperosmotic where the solution has a higher osmotic pressure than the cell (or cell organelle) or hypo-osmotic where the osmotic pressure is lower than the cell (or cell organelle).

Lipophilic   literally 'lipid- or fat-loving' (cf. lipophobic = lipid hating).

CELL MEMBRANES

Myelin sheath   concentric layers of lipid (plus some protein) surrounding the axons of certain nerve cells (neurons). The insulating layers are derived from special Schwann cells.

Opsonin   proteinaceous, cationic antibody.

Osmosis   selective diffusion of a solvent (usually water) through a semi-permeable or selectively permeable membrane.

Phagocytosis   cell-eating, i.e. cells such as *Amoeba* and white blood cells (leucocytes) ingest large particles in this way.

Pinocytosis   cell-drinking – bulk intake of fluid with large molecules.

Protozoa   unicellular or acellular animals, forming a major group (or phylum) within the animal kingdom.

Saturated fatty acids   fatty acids in which all the carbon atoms are linked by single bonds (cf. unsaturated fatty acids, in which some double bonds between carbon atoms are still present).

Selectively permeable   of biological membranes which will allow the passage of some molecules or ions across them, but not others.

Units of measurement   cm (centimetre) $= 10^{-2}$ m
mm (millimetre) $= 10^{-3}$ m
μm (micrometre) $= 10^{-6}$ m
nm (nanometre) $= 10^{-9}$ m

5.5. Appendix. Practical experiments

Experiment 1. Observations on the fate of food vacuoles in *Paramecium*

Materials
You will require the following:
    light microscope with $\times 10$ and $\times 40$ objectives
    culture of *Paramecium*
    Pasteur pipettes and bulbs (teats)
    plain slides and coverslips or cavity slides and coverslips
    petroleum jelly
    10% methyl cellulose solution or 1% agar solution
    light liquid paraffin
    access to a centrifuge
    bakers' yeast or milk
    Congo red dye

Construction of the observation chamber
This feature is common to experiments 1 and 3.
Either (*a*)   Stick 2 coverslips onto a slide with a small quantity of petroleum jelly, leaving a gap of about 1 cm between the two to form supports so that the Protozoa are not compressed.
            Place a drop of liquid paraffin on another coverslip so that it covers it, but does not flow off the edges.
            Transfer a drop of medium containing a number of Protozoa under investigation to the centre of the drop of paraffin with a drawn-out pipette (Pasteur pipette), so that it spreads on the glass surface. With fast-moving Protozoa like *Paramecium*, you will need to add 1 drop of saturated or 10% methyl cellulose solution, or alternatively 1 drop of 1% melted agar solution, to slow down their movements. Leave for 3–4 min. Now invert the coverslip and place it on the observation chamber so that the culture drop comes between your two fixed coverslips. *Avoid spilling liquid paraffin on your microscope and optics. Carefully wipe up any spillage immediately.*
or   (*b*)   Use a cavity slide and omit instructions in the first paragraph of (*a*) above. This method is not suitable for experiment 2 where irrigation of the observation chamber is necessary.

Observations on Paramecium
Add a few drops of Congo red dye (conc. 1 part : 1000 parts of absolute alcohol) to about 5 ml of milk or a suspension of yeast. Allow the dye time to mix and to be taken up by the suspension. Concentrate a sample of *Paramecium* by briefly centrifuging in a bench centrifuge. Add one or two drops of the dyed milk or yeast suspension and leave for 2 h. Food is not ingested in *Paramecium* over the whole body surface; it is only ingested at the base of the oval groove, a specialized area found in ciliated Protozoa (see Book 4 in this series).

After 2 hours observe the stained particles ingested by the *Paramecium* under the light microscope. In neutral or weakly alkaline conditions the dye is a red-coloured salt; under weakly acidic conditions it is blue.
Note. Do *not* use very intense illumination with your microscope.

Remember that the point of suggesting this experiment was to answer the question posed in frame 96; 'What happens to the engulfed body (food particle or bacterium) once entrapped in a cytoplasmic vesicle?'

Questions
1. Are all the ingested food particles the same colour or do different food vacuoles show different colours?
2. Is there any recognizable distribution pattern of food vacuoles in this organism?
3. If so, can you comment on its significance?

Experiment 2. Observations on contractile vacuole activity in relation to the concentration of the external medium in *Podophrya* (or *Paramecium*)

Hypothesis
That the function of contractile vacuoles in Protozoa is to maintain the internal solute concentration of the animal above that of its external environment, i.e. it is a means of osmoregulating (section 5.3).

Materials
You will require the following:
  light microscope
  fine forceps
  petroleum jelly
  filter paper
  Pasteur pipettes and bulbs
  clock
  culture of *Podophrya*, which has been allowed to settle on silk threads
  2%, 4% and 8% solutions of seawater

Construction of the observation chamber
Stick two coverslips on a slide with a small quantity of petroleum jelly, leaving a gap of about 1 cm between the two to form supports so that the protozoans are not compressed. Such an arrangement also allows for irrigation of the chamber.
  Mount the organism on the glass slide between the two supporting coverslips as shown in the diagram. The cotton is used as an anchor for the silk threads with attached *Podophrya*. Just loop the threads around the cotton carefully using fine forceps.

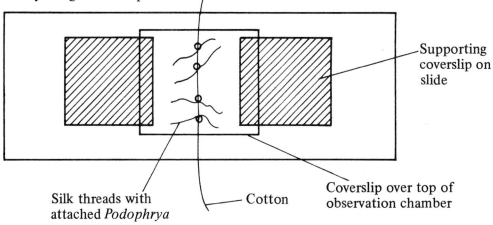

Supporting coverslip on slide

Silk threads with attached *Podophrya*

Cotton

Coverslip over top of observation chamber

EXPERIMENTS

Allow time for the organism to settle, then observe (fairly low illumination) the cycles of contractile vacuole activity under the light microscope. Using a continuously running clock record the number of vacuolar cycles, say per 5 min. If your microscope has a calibrated eyepiece (see Book 1 in this series) you may like to estimate the diameter (d) of the contractile vacuole just before it collapses. The mean vacuolar volume can then be estimated from the formula

$$v = \frac{d^3}{6}.$$

Now irrigate with 2% seawater by slightly tilting the slide as shown below.

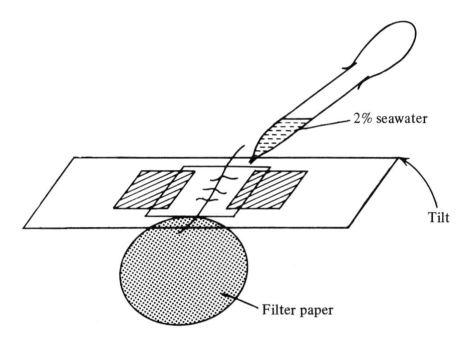

Repeat your observations. Repeat again, using 4% and then 8% seawater, finally returning the organism to tapwater.

Plot graphically the mean rate of fluid output ($\mu m^3/s$) against concentration (%) of seawater.

i.e.

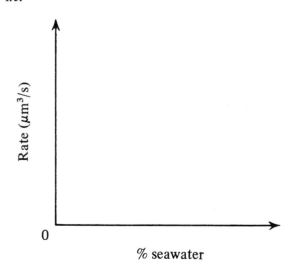

CELL MEMBRANES

Do your results support the hypothesis presented at the beginning of this experiment?

Experiment 3. Observations on the chemotactic behaviour of amoebae

Materials
You will require the following:
  light microscope
  culture of *Amoeba*
  Pasteur pipettes and bulbs
  plain slides and coverslips or cavity slides and coverslips
  petroleum jelly
  light liquid paraffin
  culture of *Hydra*
  anionic and cationic exchange resins

Construction of observation chamber
Set up as in experiment 1. You do not need 10% methyl cellulose or 1% agar solution for observing amoebae.

Observations on amoebae
When observing the behaviour of *Amoeba*, do *not* use intense illumination with your microscope.
1. Cut up a hydra into small pieces and put a piece into a depression slide with amoebae. Carefully observe your preparation over a period of at least 15 min and record what happens. Over what distance are amoebae affected by hydra?
2. Repeat experiment 1 using small particles of cation exchange resin (negatively charged) in one preparation and anion exchange resin (positively charged) in the second. Observe the behaviour of the amoebae.
Remember that the point of this experiment is to provide evidence for the mechanism whereby amoebae (and possibly amoeboid cells) are attracted to specific chemical stimuli.

Questions
On the basis of the above experiments what conclusions can you draw about chemotaxis in amoebae?
What is the nature of the chemotactic stimulus?
What is the significance of the amoeba's behaviour and what role does it play in the animal's life?

5.6. Questions relating to the objectives of the book

The following questions may be used as pre and post test to this book or alternatively as self-assessment questions. The objective(s) which is tested in each case is indicated in brackets after each question.

Cell membranes and transport

1 Label the following compounds 1, 2, 3, 4, 5 and 6 according to their rate of *diffusion* through a plasma membrane. Use 1 for the fastest and 6 for the slowest:

Glucose
Water
Sucrose
Potassium ion
Oxygen gas
Serum albumin

(*Objectives 1 & 4*)

2 Which of the following observations or pieces of experimental evidence are compatible, or might be compatible (C) with the unit membrane structure as proposed by Robertson, and which are incompatible (I)?

(i) The electrical properties and permeability of plasma membranes to small ions and water soluble molecules can be explained by 'pores' which allow free diffusion of polar molecules.

(ii) Freeze-etched electron micrographs of myelin membranes.

(iii) Active transport, carrier mechanisms and hormonal receptor sites.

(iv) Biochemical evidence indicates that polar mechanisms do not dominate the interactions between proteins and lipids.

(v) Freeze-etched electron micrographs of plasma membranes in general.

(vi) When lipids extracted from erythrocyte membranes are spread evenly as a monolayer on water in a Langmuir trough, they occupy an area equivalent to twice the area of the cells from which they were extracted.

(vii) Optical measurements indicate that most of the protein is in globular form.

(viii) The trilaminar appearance of the inner mitochondrial membrane remains essentially unchanged, even after removal of 95% of the membrane lipid, when permanganate is used as fixative for electron microscopy.

(ix) The tension at the surfaces of certain cells is lower than that at lipid−water (or oil−water) interfaces, yet adsorption of denatured protein to oil droplets markedly lowers their interfacial tension.

(*Objectives 2 & 4*)

3 The following is a model of a membrane as proposed by Lucy (1964):

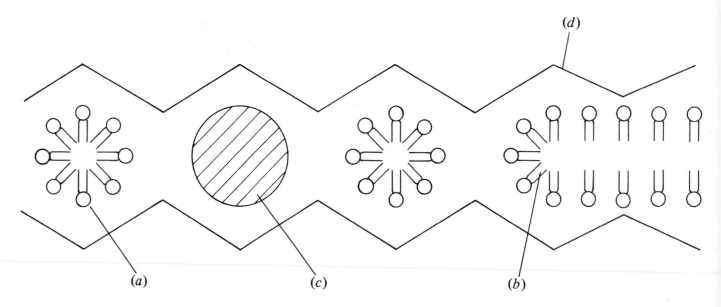

(a)   is the _____ end of a _____ molecule
(b)   is the _____ end of a _____ molecule
(c)   is _____
(d)   is _____
Fill each blank with 1 or 2 of the following words:
polar, non-polar, structural, functional, protein, phospholipid, glycoprotein

(*Objective 2*)

4   If you place a red blood cell in a strong sugar solution which of the
      following would you expect to happen? Choose one alternative from
      each set.
      (i)   *a.*   a large flow of water out of the cell
            b.   a small flow of water out of the cell
            c.   a large flow of water into the cell
            d.   a small flow of water into the cell
      (ii) *a.*   a large flow of sugar out of the cell
            b.   a small flow of sugar out of the cell
            c.   a large flow of sugar into the cell
            d.   a small flow of sugar into the cell
      (iii) *a.*   expansion of the cell
            b.   no change in the size of the cell
            c.   shrinking of the cell
            d.   haemolysis of the cell

(*Objective 1*)

5   The statements below refer to changes in glucose concentration in
      hamster gut. The initial concentration of glucose was the same on both
      sides of the gut. Mark the statement O if it is true under aerobic
      conditions (oxygen present), N if it is true under anaerobic conditions,
      B if it is true for both, and X if it is true for neither.
      (a)   Glucose diffuses from the inside (lumen) to the outside.
      (b)   Glucose diffuses from the outside to the inside.
      (c)   Glucose is transferred from the inside to the outside by active
            transport.

     (*d*)   Changes in glucose concentration are dependent on the presence of Na^+ ions.

     (*e*)   Lactic acid is produced.

(Objectives 1 & 5)

6   In red blood cells diffusional tendencies lead to a net outward diffusion of _____ ions and a net inward diffusion of _____ ions; active transport results in an outward movement of _____ ions and an inward movement of _____ ions. (Fill in blanks with Na^+ or K^+.)

(Objective 1)

7   The following changes have been observed in the Na^+ and K^+ concentrations inside red blood cells over a period of 24 h.

	Na⁺ inside	*K⁺ inside*
Initially	25 mM	140 mM
After 24 h	110 mM	20 mM

Which of the following might be true, (mark P for possible), must be true (mark T), or must be false (mark F)?

     (*a*)   The cells are dead and have been kept in 150 mM NaCl at 37 °C.

     (*b*)   The cells have been kept in 130 mM NaCl + 20 mM KCl + glucose at 37°C.

     (*c*)   The cells have been kept in 150 mM NaCl + glucose at 2°C.

     (*d*)   The cells have been kept in 130 mM NaCl + 20 mM KCl at 37 °C.

     (*e*)   The cells have been kept in 250 mM NaCl + glucose at 37 °C.

(Objectives 3 & 5)

8   Which of the following processes connected with active transport in red blood cells are thought to occur at the same rate? (Choose more than one.)

     (*a*)   Breakdown of glucose

     (*b*)   Movement of K^+ ions

     (*c*)   Movement of Na^+ ions

     (*d*)   Breakdown of ATP

(Objectives 1 & 3)

9   Which of the following are behavioural or structural features of white blood cells? (Choose more than one.)

     (i)   Have a nucleus

     (ii)   Show intrinsic movements

     (iii)   Have heterogeneous cytoplasmic inclusions

     (iv)   Are frequently phagocytic

     (v)   Transport oxygen in the blood

     (vi)   Have a regular shape

(Objective 6)

10   Phagocytosis is elicited by certain stimuli. Which of the following would give a positive phagocytic response and which would not? Indicate your answer by writing P for always a positive result, N for always a negative result, and S for sometimes a positive result.

     (i)   Chemical stimulation by substances bearing a positive charge

     (ii)   Chemical stimulation by substances bearing a negative charge

75

(iii) Chemical stimulation by substances bearing no charge
(iv) Physical contact with a number of substances
(v) Physical contact with neutral substances complexed with cationically charged particles.

(Objective 6)

11 Which of the following are features of opsonins? (Choose more than one.)
(i) opsonins are proteins
(ii) opsonins are antibodies
(iii) opsonins are antigens
(iv) opsonins are anionic substances
(v) opsonins change the surface membrane properties of bacteria

(Objective 6)

12 The hypothesis that certain living cells can release a chemical or chemicals which will attract amoeboid cells was tested by Francis (1965) by randomly placing a group of amoeboid cells in a flow chamber with a small mass of animal cells.
Mark the following statements about the experiment as either *true* or *false*.
(*a*) The following graph of results obtained after one hour's observation shows that the mass of cells is responsible for the distribution of amoeboid cells.

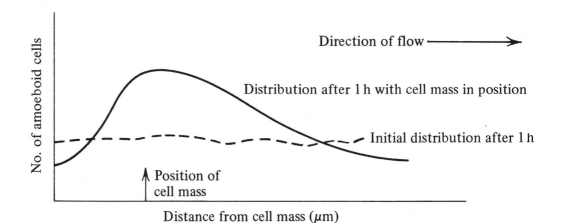

(*b*) The experimental results in (*a*) indicate that the distribution of amoeboid cells is determined merely by the flow of medium through the observation chamber.
(*c*) The experimental results in (*a*) precludes the possibility that attraction is caused by the influence of an electrical field.
(*d*) A control experiment in which there is no continuous flow of medium through the chamber is essential.
(*e*) A control experiment is not required.

(Objective 6)

Answers

1   (1) oxygen, (2) water, (3) glucose, (4) potassium ion, (5) sucrose, (6) serum albumin.

2   (i) C, (ii) C, (iii) C, (iv) I, (v) I, (vi) I, (vii) I, (viii) I, (ix) C.

3   (a) polar, phospholipid; (b) non-polar, phospholipid; (c) functional protein; (d) structural glycoprotein.

4   (i) a, (ii) d, (iii) c.

5   (a) B, (b) B, (c) O, (d) O, (e) N.

6   K^+, Na^+; Na^+ K^+.

7   (a) P, (b) F, (c) T, (d) P, (e) F.

8   b, c, d.

9   v, vi.

10   (i) P, (ii) N, (iii) N, (iv) S, (v) P.

11   i, ii and v.

12   (a) True, (b) False, (c) True, (d) False, (e) True.

5.7. Recommended reading

Branton, D. (1969)   Membrane structure. *Ann. Rev. Plant Physiol.* **20.**
                          209.
Branton, D. (1971)   *Freeze-etching studies of membrane structure.*
                          Phil. Trans. Roy. Soc. B. **261**, 121.
Davies, M. (1973)   *Functions of biological membranes.*
                          Outline Studies in Biology. Chapman & Hall, London.
Edelman, G.M. (1970)   The structure and function of antibodies.
                          Sci. Am. (October) p. 1185.
Finean, J.B. (1972)   The development of ideas on membrane structure.
                          Sub-Cell. Biochem. **1**, 363.
Lockwood, A.P.M.   *The membranes of animal cells.*
  (1970)   Studies in Biology no. 27. Edward Arnold, London.
Nossal, G.J. (1964)   How cells make antibodies.
                          Sci. Am. (December) p. 199.
Pastan, I. (1973)   Cyclic AMP. *Sci. Am.* (August) p. 97.
Reisfeld, R.A. &   Makers of biological individuality.
  Kahan, B.D. (1972)   *Sci. Am.* (June) p. 28.
Spiers, R.S. (1964)   How cells attack antigens.
                          Sci. Am. (February) p. 58.
Stein, W.D. (1969)   *The movement of molecules across cell membranes.*
                          Academic Press, New York.
Wallach, D.F.H.   *The plasma membrane: dynamic perspectives,*
  (1972)   *genetics and pathology.* Heidelberg Science
                          Library, vol. 18. Springer-Verlag, Berlin,
                          Heidelberg & New York.

Index

active transport, 30–48, 64
adenosine triphosphate (ATP) 44–8, 65
aerobic respiration 33–5, 67
amino acids 38
Amoeba 67, 72
anaerobic respiration 33–5, 67
antibody 58, 67
antigen 58, 67
artificial membranes 13

bacteria 51–3
Baldridge & Gerrard 64
blood groups 66
Branton 22

chloroplast lamellae 20, 21

Danielli & Davson 11
diffusion 26, 68
diffusion coefficient 66

endoplasmic reticulum 21

facilitated diffusion 39, 68
Finean 21
Francis 53
freeze-etched preparations 19, 68

glucose 31–7, 42–4
Gorter & Grendel 8

hydrophobic 19, 68

intestine 29, 68

lactic acid 35, 36
Landsteiner 67
lipophilic 68
Lucy 16, 24, 25

mitochondrial membranes 14, 15, 18
mucosal side of intestine 32, 33
myelin sheath 18, 20

non-polar (= apolar) 5, 6
nuclear membrane 18

opsonin 58
osmium tetroxide (OsO_4) 4, 11
osmoregulation 65
osmosis 28–30, 68
Overton 4, 65

Paramecium 70–3
phagocytosis 53–7, 65
pinocytosis 61–5
Podophrya 70, 71
polar 6–10
'pores' in membranes 24, 25
Post & Jolly 40
potassium permanganate ($KMnO_4$) 4, 11

red blood cell (= erythrocyte) 12, 20, 23, 39–48,
    51, 52
Robertson 11

saturated fatty acids 6
Schultz 37
selective permeability (of membranes) 5, 25, 26,
    69
serosal side of intestine 32, 33
Sjöstrand 16
sodium-dependent amino acid uptake 38
sodium-dependent glucose uptake 36, 37
sodium pump 40–4, 65
surface area to volume ratio 27, 28

unit membrane hypothesis (= theory) 8, 11, 24
unmyelinated nerves 12
unsaturated fatty acids 6

Wallach 18
white blood cells (= leucocytes or phagocytes)
    51, 52, 56
Whittam 43
Whittam & Ager 46
Wilson & Wiseman 32, 35, 36

Zimmerman & Rustad 64